G. Contopoulos   D. Kotsakis

Cosmology

The Structure and Evolution of the Universe

Translated by M. Petrou and P. L. Palmer

With 54 Figures

Springer-Verlag Berlin Heidelberg New York
London Paris Tokyo

Professor Dr. *Georgios Contopoulos*
Astronomy Department, University of Athens, Panepistimiopolis
GR-15783 Zografos, Athens, Greece

Professor Dr. *Dimitrios Kotsakis* †

Translators:

Dr. *M. Petrou*
NERC Unit for Thematic Information Systems, Department of Geography, White Knights
P.O. Box 227, IV. Reading, Berks, RG 6 2BA, England

Dr. *P. L. Palmer*
Queen Mary College, University of London, Mile End Road
London E1 4NS, England

Cover picture: The supernova remnant in Vela
(Photograph: European Southern Observatory)

Title of the original Greek edition:

Kosmologia – I Domi Kai I Exelixi Tu Sympantos, 2nd updated and expanded edition
© G. Contopoulos and D. Kotsakis, Athens 1984

ISBN 3-540-16922-9 Springer-Verlag Berlin Heidelberg New York
ISBN 0-387-16922-9 Springer-Verlag New York Berlin Heidelberg

Library of Congress Cataloging-in-Publication Data. Kontopoulos, Geōrgios Iōannou, 1928 – Cosmology : the structure and evolution of the universe, with 54 figures. Translation of: Kosmologia. 1. Cosmology. I. Kotsakis, D. (Dimitrios). II. Title. III. Title: Structure and evolution of the universe. QB981.K69513 1987 523.1 86-21920

© Springer-Verlag Berlin Heidelberg 1987
Printed in Germany

The use of registered names, trademarks, etc. in this publication does not imply, even in the absence of a specific statement, that such names are exempt from the relevant protective laws and regulations and therefore free for general use.

Media conversion, offsetprinting and bookbinding: Graphischer Betrieb Konrad Triltsch, Würzburg
2153/3150-543210

Preface

This is a translation from the second Greek edition (1984). We have added a few more sections and increased the bibliography to keep up with the developments in cosmology over recent years. In this edition, the parts containing mathematical supplements and various digressions are indicated by a vertical grey line in the margin and may be skipped if desired.

We wish to thank Drs. M. Petrou and P. Palmer for the translation of the book. We also thank Dr. D. Kazanas for his remarks.

May 1986   *G. Contopoulos · D. Kotsakis*

Preface to the First Greek Edition

This book has been written in close collaboration between the two authors. However, D. K. is mainly responsible for the first part, while G. C. is mainly resonsible for the 2nd and 3rd parts. The bibliography has been updated up to the beginning of 1982.

We thank our colleague Dr. B. Barbanis for many useful comments on the book, Drs. P. Mihailidis and A. Pinotsis for drawing the figures and Dr. M. Zikidis for preparing the cover page.

May 1982   *G. Contopoulos · D. Kotsakis*

Obituary

In May 1986 Professor D. Kotsakis passed away.

Dimitrios Kotsakis was born in Filiatra, Greece, in 1903. He graduated from the University of Athens in 1931 and became Professor of Astronomy in 1965. His scientific work was both observational and theoretical. Among other things he studied the dispersion of the fragments of a hypothetical exploding planet. He also studied pure mathematical problems in the field of differential equations etc., and the history of science in Greece. He published more than 20 books in Greek, for students and the general public, which had considerable success.

His most important contribution to Greek astronomy was the establishment of the new observatory in Kryonerion, near Corinth. The Kryonerion Observatory, which contains a 120-cm telescope, the largest in Greece, belongs to the National Observatory and has been in operation since 1975. D. Kotsakis also encouraged and supported several students to study astronomy abroad. Thus D. Kotsakis will be remembered as one of the main individuals who brought about the modernisation of Greek astronomy.

Contents

Part II Theory

Part III Fundamental Problems

Introduction

Cosmology is a relatively new science. Its beginning must be placed around 1929, when *Hubble* discovered the expansion of the Universe. The most amazing thing about this discovery is the universality of the expansion. *All* the galaxies are going away from us, and actually with velocities which increase with their distance. Therefore, this is neither a local phenomenon nor a random statistical event. The *whole* Universe expands, all the galaxies go away from each other with enormous velocities which, at large distances, approach the speed of light. The discovery verified some of the most daring predictions of Einstein's general theory of relativity.

The general theory of relativity was published in 1916 and its first application to our solar system had already been made before 1920. But only very few bold mathematicians and astronomers had the idea of extending its consequences to the whole Universe. The first people to construct models of an expanding Universe were *de Sitter* (1917), *Friedmann* (1922) and mainly *Lemaitre* (1927). However, the majority of astronomers did not take these models seriously. Only *Eddington* thought, in 1922, that the de Sitter model could represent the real Universe of galaxies. Similar ideas were expressed by *Lemaitre*. This situation lasted until Hubble's discovery in 1929. The whole way of thinking was altered after that. For the first time the study of the Universe as a whole stopped being the subject of personal pondering and became the subject of scientific research. The great advancement of cosmology that followed was due to systematic research in observations and theory. The phenomenon, however, of the expansion of the Universe was so enormous, so amazing, that many people disputed it. Many efforts were made to attribute the redshift of the light from distant galaxies to causes other thant the expansion of the Universe. All these efforts failed and today there is no serious dispute of the reality of the expansion.

Hubble initiated a large scale study of the Universe, starting from the nearby galaxies. The galaxies are the basic ingredients of the Universe, its "atoms". This study was aided by a new generation of large telescopes, first being the 5 m-telescope at Palomar. It was followed by the 4 m-telescope at Kitt Peak, USA, the 6 m-telescope in Caucasus, USSR, and the three largest telescopes in the southern hemisphere, the 4 m-American in Chile, the 3.6 m of the "European Southern Observatory" also in Chile, and the 3.9 m-Anglo-Australian telescope in Australia.

Also, radio astronomy gave new impetus to the study of the Universe up to very large distances. In particular, the discovery of quasars, which was first made by radio telescopes, allowed us to expand our search up to distances of about 10 billion light years.

In recent years, the study of x- and γ-rays from space, as well as the microwave, infrared and ultraviolet radiation opened up new ways of exploring the Universe. The greatest discovery in this area was the microwave background radiation. *Penzias* and *Wilson* who discovered it in 1965 received the Nobel prize in physics in 1978. This radiation, which comes uniformly from all points in the Universe, corresponds to the radiation of a black body at temperature 3 K, i.e. 3 degrees above absolute zero. The only plausible explanation for the origin of this radiation is that it is the remnant of an early phase in the expansion of the Universe. The photons of that phase spread throughout space, and lost their energy through the cosmic expansion, so that their corresponding temperature now is only 3 K. It is interesting to note that the existence of this radiation had already been predicted in 1946–48 by *Gamow* and his collaborators, as a consequence of the Big Bang theory.

The discovery of the microwave background radiation was the greatest factor in the evolution of cosmology in recent years. Of course a lot of attempts were made to attribute this radiation to stars or galaxies or other sources. However, two of its properties confirm its cosmological origin. First, its isotropy, i.e. the fact that its intensity is the same in all directions; and second the confirmation of its black-body nature.

One main conclusion, after many years of research in cosmology, is that one should take very seriously the general theory of relativity and its predictions. Many other theories formulated at various times in order to explain the Universe, have been abandoned. Thus, the interest in general relativity has grown spectacularly in the past 20 years. Among the major contributors to the revival of this theory are *Bondi, Chandrasekhar, Kerr, Hawking, Penrose* and others. In recent years, thousands of papers have been published on relativity and cosmology.

The most important characteristic of modern cosmology is the involvement of high-energy physicists in it. Among them, *Weinberg, Salam, Glashow, Cronin, Fitch, Rubbia* and *van der Meer* received Nobel prizes in recent years (1979–1984) for their research which is relevant to cosmology. Thus, a very serious effort is being made today to unify general relativity with the theories of microphysics, so that one theory should describe both the phenomena of the microcosm and of the whole Universe.

As early as 1946 *Gamow* and his collaborators had started the theoretical research for the origin of the various elements during the first four minutes of the expansion of the Universe. Recently, this research has been extended to cover the origin of the elementary particles themselves. Thus, physicists have started to investigate the conditions in the Universe immediately after the initial explosion, when its age was a small fraction of a second, its temperature trillions of degrees and the radius of the presently visible Universe less than the radius of an atom!

Cosmology today is a whole science by itself. It includes theory and observations, just like any other science. It has a close relationship with many other sciences and uses their results. However, it has also its own peculiar characteristics and methods, which we shall describe in this book.

In general, cosmology is much wider now than it was 20 years ago. Not only does it touch a wide spectrum of problems, but it has entered into the deepest problems of physics and tries to give an answer to the most fundamental questions of physics and astrophysics. The award of a Nobel prize of physics (1983) to the astrophysicists *Chandrasekhar* and *Fowler* is a recognition of the unification of the problems of physics and astrophysics.

A proof of the current activity in cosmology is the number of relevant papers published and meetings held. Hundreds of scientific papers and several books are published every year on cosmological subjects. At the end of this book we mention more than 70 books on cosmology published since 1970. Also, a lot of meetings on cosmology are held all over the world. We shall mention here only some relevant ones organised by the International Astronomical Union. The International Astronomical Union, which includes the astronomical activity worldwide, organises every year symposia and colloquia on current subjects, where specialists from all over the world meet to discuss them. More specifically, after 1970 the following meetings related to cosmology have been organised:

1. Symposium 58, "The Formation and Dynamics of Galaxies", Canberra, Australia, 1973.
2. Symposium 63, "Confrontation of Cosmological Theories with Observational Data", Cracow, Poland, 1973.
3. Symposium 64, "Gravitational Radiation and Gravitational Collapse", Warsaw, Poland, 1973.
4. Symposium 74, "Radioastronomy and Cosmology", Cambridge, England, 1976.
5. Colloquium 37, "Redshift and the Expansion of the Universe", Paris, France, 1976.
6. Symposium 79, "The Large Scale Structure of the Universe", Tallinn, USSR, 1978.
7. Symposium 92, "Objects of High Redshift", Los Angeles, USA, 1979.
8. Symposium 97, "Extragalactic Radio Sources", Albuquerque, USA, 1981.
9. Symposium 104, "Early Evolution of the Universe and its Present Structure", Kolymbari, Crete, Greece, 1982.

10. Symposium 117, "Dark Matter in the Universe", Princeton, USA, 1985.
11. Symposium 124, "Observational Cosmology", Shanghai, China, 1986.

Cosmological problems are also discussed in many other conferences of the International Astronomical Union and other international and national unions (like the European Physical Society, the American Astronomical Society etc).

There is no up-to-date book on cosmology in the Greek bibliography appropriate for a wider reading public. That is why we considered it appropriate to proceed in publishing this book which contains the current developments in cosmology.

The book is written, in general, with very little mathematics. At some points, however, it was necessary to use advanced concepts of physics or mathematics. In these cases we try to explain these concepts as simply as possible. In order to make things easier for the reader, we mark the sections which require particular knowledge, or more difficult mathematics, by a vertical grey line in the margin so that the reader may omit them on first reading.

Of course, in a restricted work on cosmology, like the present one, it is not possible to include everything written in the thousands of related scientific papers published in recent years. So, quite often we briefly describe conclusions reached after research which could fill up whole books in itself. But, in order to offer the opportunity to those interested to study in greater detail particular cosmological subjects, we include, at the end of this book, a list of the most important related books and of some recent interesting papers in the field.

Part I

Observations

1. Our Galaxy

1.1 Observing Our Galaxy

The stars are not uniformly distributed in space but are segregated in galaxies separated by almost empty space. The Sun and its neighbouring stars belong to one such galaxy which looks like a lens. This is *our Galaxy*. In this chapter we are going to discuss what we know about it.

On summer nights we can actually see our Galaxy in the sky. It looks like a milky strip on the celestial sphere, which is why it is called the "Milky Way" ($\Gamma\alpha\lambda\alpha\xi\iota\alpha\varsigma$, in Greek). It is particularly impressive if we view it from the countryside, far from the city lights.

The central line along this strip coincides with what is called the galactic plane (Fig. 1.1). Away from this plane the density of stars drops quickly, but the boundaries of the milky strip are rather diffuse. On the other hand the density of stars increases as we approach the centre of the Galaxy, which is brighter than any other part we see.

Since we see our Galaxy from within, it is very difficult to determine its shape. As far as we can tell it is lens-shaped (Fig. 1.2). However, if we were able to observe it from a distance along its symmetry axis we would see that it looks like a spiral (Fig. 1.3).

Let us try to count the stars of our Galaxy. The brightness of a star is measured in magnitudes. The brightest stars are of the 1st magnitude and the faintest stars observable by the naked eye are of the 6th magnitude. (The fainter the star, the larger its magnitude.) Suppose that $N(m)$ is the number of stars brighter than magnitude m. If space were uniformly filled with stars and if we could see stars infinitely far away, it can be shown that $N(m + 1)/N(m) = 3.98$ (Sect. 3.5). The number we deduce from actual star counts, however, is much smaller than this. There are two reasons for this: (1) The density of stars drops dramatically in the outer parts of the Galaxy, thus space is not uniformly filled with stars. (2) There is a lot of interstellar dust, which makes the stars look even dimmer than they really are.

The number of stars we observe decreases as we observe to larger angular distances from the galactic centre. This is true not only if we

---

Fig. 1.1. Our Galaxy as seen by an observer on the Earth (Lund Observatory, Sweden)

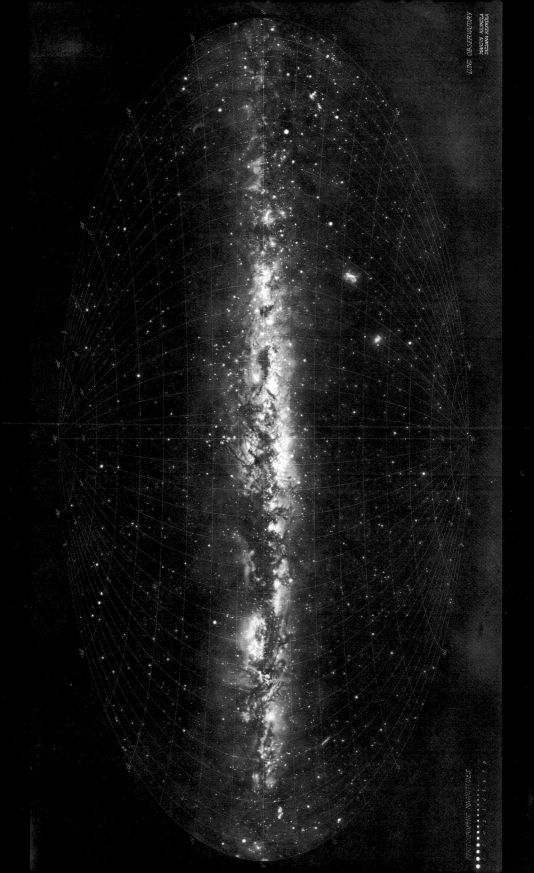

PHOTOGRAPHIC MAGNITUDES

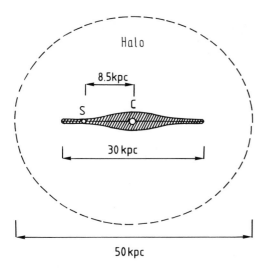

Fig. 1.2. Our Galaxy is lens-shaped and embedded in a halo which has a diameter of 50 kpc (kiloparsecs). The diameter of our Galaxy is 30 kpc and our Sun, S, is 8.5 kpc away from the galactic centre, C

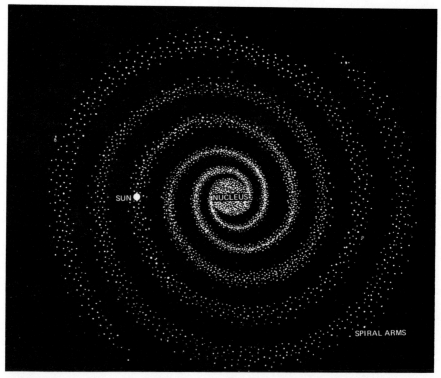

Fig. 1.3. The spiral structure of the Galaxy as seen from an observer on its axis of symmetry

observe away from the galactic plane, but inside the plane also. More specifically, the number of faint stars, which presumably are the more distant ones, changes along the Milky Way. There is a maximum of the density of stars in the galactic plane in the direction of the constellation of Sagittarius, while there is a minimum in the opposite direction towards Auriga. This implies that the Sun is not at the centre of the Galaxy but rather far from it. It has been estimated that the distance of the Sun from the galactic centre is about 8500 pc[1] (Figs. 1.2, 1.3). There are more than 1 000 000 000 000 stars in our Galaxy. Some of them are grouped together in *globular clusters,* in *open* (or *galactic*) *clusters,* or in *associations.* Apart from stars there are also *bright and dark nebulae* of interstellar medium. The interstellar space is also filled up with *cosmic rays* and radiation emitted from the stars in all wavelengths: *radio waves; infrared, optical and ultraviolet radiation; x-rays and γ-rays.*

1.2 The Hertzsprung-Russell Diagram

The Hertzsprung-Russell (H-R) diagram is a plot of the luminosities versus the temperatures of various stars. Stars of different temperatures radiate at different wavelengths with different intensity, so there is a great variety of stellar spectra. They can, however, be classified into seven categories which in order of descending temperature are: O, B, A, F, G, K, M. At the end of this sequence one may add the carbon stars (C) and the zirconium stars (S). Stars of these latter categories are of very low temperature, less then 3000 K.[2]

We say that stars of spectral type O or B are of early type, while those with spectral types C, K, M or S are of late type. The transition from one spectral type to the next is gradual and we distinguish 10 intermediate types between any two successive main categories (e.g. we have stars B2, M8 etc.). Almost all the stars belong to one of the above spectral types.

The spectral types can be used as the abscissae in the H-R diagram instead of the temperatures (Fig. 1.4). An alternative quantity which may also be plotted along the abscissa is the colour index of a star defined as the difference between its photographic and optical magnitudes. The greater the temperature of the star, the smaller its colour index.

In the H-R diagram most of the stars fall along a line from the top left to the bottom right of the diagram, called the *Main Sequence*. There are also small hot stars called white dwarfs plotted in the bottom left of the

---

[1] The parsec (pc), as well as the light year (ly), are units used for measuring large distances. The following relations hold: 1 pc = 30.857×10^{12} km = 3.262 ly; 1 kiloparsec (kpc) = 1000 pc; 1 Megaparsec (Mpc) = 1000 kpc = 10^6 pc; 1 ly = 9.4605×10^{12} km = 0.3066 pc.
[2] These are degrees Kelvin measured from the absolute zero (-273 degrees Celsius).

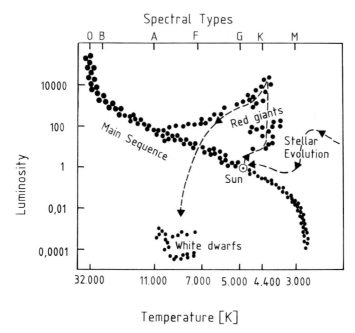

Fig. 1.4. The Hertzsprung-Russell (H-R) diagram. Along the abscissa we mark the spectral types (*top*) and the surface temperatures (*bottom*) of the stars in degrees Kelvin, and along the ordinate the absolute luminosities in solar units. Most of the stars are in three areas: the Main Sequence, the red giants and the white dwarfs. The mass and the radius of a star increase along the Main Sequence from the bottom right to the top left. (---) represents the evolutionary path of the Sun. The present position of the Sun is marked by ⊙

diagram as well as large cold stars called red giants plotted at the top right. The gap appearing between the Main Sequence and the red giants is called the "Hertzsprung gap".

Figure 1.4 shows the H-R diagram for the stars in the solar neighbourhood. A star, during its lifetime is expected to change position in the diagram. The dashed line on Fig. 1.4 represents the "evolutionary track" for a star like the Sun. After its birth the star is bright and red having just condensed out of the interstellar medium. The star soon reaches the Main Sequence where it spends most of its life. It will later evolve to become a red giant and finally a white dwarf. When a star is on the Main Sequence we say that it has zero age. During this phase of its life the star produces energy by converting hydrogen into helium in its nucleus. As the hydrogen becomes exhausted the star moves away from the Main Sequence. The red giants owe their name to their large size and red colour. They are very bright but their surface is relatively cool. In the nucleus, however, the temperature is high enough for helium to start being converted into other elements. When our Sun becomes a red giant its diameter will be 1000 times its present size, with the inner planets Mercury and Venus buried inside it. It is estimated that this will happen in about 5×10^9 years time. Eventually as the fuel in the solar nucleus burns out, the Sun will contract to a white dwarf.

The more massive a star is, the faster it evolves. Stars with masses less than 1.4 solar masses $(M_\odot)$ evolve in the same way as the Sun. Stars with higher masses quickly cross the Hertzsprung gap to become red giants and finally evolve to either "neutron stars", if their masses are in the range $1.4-3\ M_\odot$, or "black holes" if their masses are greater than $3\ M_\odot$ (Sect. 5.8).

1.3 Globular Clusters

In our Galaxy we can see groups of stars which are held together by their mutual gravitational attraction and form clusters. There is one category of clusters which are particularly rich in stars, and are called globular clusters. Figure 1.5 shows a globular cluster which consists of 300 000 stars. We can count the density of stars in the outer parts of the cluster and extrapolate towards the central regions where we cannot distinguish individual stars. In this way we find that a globular cluster may contain anything from a few hundred thousand up to a few million stars. The diameter of a globular cluster is from 10 to 100 pc and the density of stars is between 1000 and 10 000 times higher than in the solar neighbourhood. It has been estimated, for example, that there are 100 to 1000 stars per cubic parsec in the centre of a globular cluster.

As mentioned in Sect. 1.2 the position of a star in the H-R diagram depends upon its mass and its age. Suppose now that we have a group of stars of various masses but with the same age and the same initial composition. For each one of these stars we can plot its evolutionary track in the H-R diagram, using the current theory of stellar evolution. Then by connecting points reached by the various models at the same time we can draw what we call isochrone curves (Fig. 1.6).

The stars in a globular cluster are thought to have a common origin and thus common initial composition and age. We can make a direct comparison, therefore, between the theoretical isochrone curves and the H-R diagram constructed for the members of a globular cluster (Fig. 1.7). The conclusion we draw is that the globular clusters are the oldest objects in our Galaxy, being 8-16 billion years old.

There are about 125 known globular clusters in our Galaxy, but their total number is thought to be around 200. They are spherically distributed about the galactic centre up to a radius of 100 kpc, with a maximum concentration towards the galactic centre.

Fig. 1.5. Photograph of the globular cluster ω Centauri comprising more than 300 000 stars at a distance of 25 000 ly from us. The stars orbit its centre in a few million years (European Southern Observatory)

1.4 Open Clusters

The open clusters are star clusters containing only a small number of stars (a few dozens to a few thousands). Their diameters vary between 2 and 20 pc. Some of the best known open clusters are the Pleiades and the Hyades in the constellation of Taurus as well as the h and χ Persei clusters. The Pleiades cluster includes about 120 stars inside a sphere of diameter 4 pc. The average density, therefore, is 4 stars per cubic parsec, much lower than the density inside a globular cluster. Even so, it is much higher than the density in the solar neighbourhood which is only 0.07 stars per cubic parsec. Other open clusters may have higher densities.

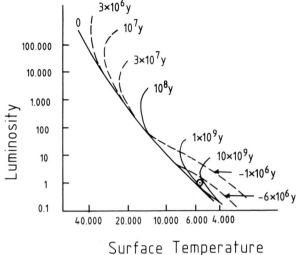

Fig. 1.6. Isochrone curves on the H-R diagram for stars of various masses before they reach the Main Sequence (negative times ---), and as they turn towards the red giant branch (positive times ——). The stars on the Main Sequence are considered to have zero age. The ages are in years (y)

When an open cluster is very close to us it may not appear as an obvious concentration of stars. Its members, however, can be distinguished from other stars projected onto the same area of the sky (called "field stars") by the fact that they all move in the same direction.

As we mentioned in the previous section, the globular clusters are of very great age and thus all their bright members have evolved away from the Main Sequence in the H-R diagram. In contrast, the H-R diagram for an open cluster consists of a well defined Main Sequence, with a branch which turns to the right and maybe some stars in the red giant region (Fig. 1.7). The branch which turns to the right is due to the more massive members of the cluster which have already evolved away from the Main Sequence. The older the cluster is, the more time its members have had to evolve and the turning point off the Main Sequence is found further to the right. For example, the cluster NGC 2362[3] consists almost entirely of Main Sequence stars. Comparing the H-R diagram for a certain cluster with a theoretically constructed one, we can deduce the age of the cluster. We find, for example, that the two clusters h and χ Persei are among the youngest, being only a few million years old. Only stars with masses

---

[3] The first catalogues of nebulae and clusters are the Messier Catalogue (M) and the New General Catalogue (NGC). They include clusters, clouds and galaxies. Thus the galaxy of Andromeda is known as M31 and also as NGC 224.

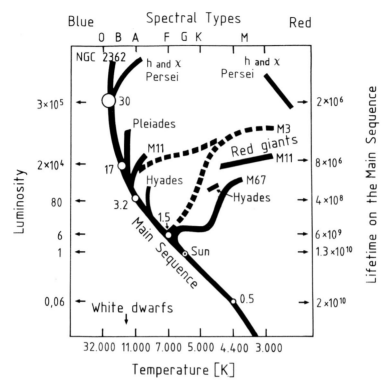

Fig. 1.7. Composite H-R diagram for various open clusters and comparison with the globular cluster M3. The numbers on the Main Sequence are the masses in solar units. The ages of the clusters for each turning point off the Main Sequence are marked on the right in years

greater than 30 $M_\odot$ have had a chance to evolve away from the Main Sequence in these clusters. Some of them have already reached the red giant stage. The Pleiades and Hyades clusters are older (20 and 400 million years old respectively). Even so, they are younger than the Earth. The oldest open cluster is about 6×10^9 years old. It is estimated that there are ten to a hundred thousand open clusters in our Galaxy.

Generally, the younger clusters have more stars, while the older ones are more diffuse due to the disassociation of their members. There are three causes which contribute to the dissolution of a cluster: (1) Encounters with massive interstellar clouds may trigger the escape of some cluster stars. (2) Encounters between stars inside the cluster cause energy exchanges which may result in some of the members acquiring a velocity in excess of the cluster "escape velocity". (3) The tidal force due to our Galaxy may cause some stars to be ejected. For the above three reasons the clusters gradually disintegrate. The average lifetime of an open cluster is of the order of 10^8–10^{10} years while that of a globular cluster is of the order of 10^{12} years.

1.5 OB Associations

These are relatively loose associations of stars of spectral type O and B, first observed by *Ambartsumian* in 1949. They are usually found along the spiral arms of our Galaxy (e.g. the ζ Persei association). Their diameter is about 100 pc but their density is no higher than the average stellar density in their neighbourhood of the Galaxy. They are distinguished from the other stars in the same area by their spectra. Observations at radio wavelengths reveal that they are surrounded by neutral hydrogen.

OB associations are composed of very young stars, some of them no older than a few million years. *Blaauw* in 1952, obseved that the ζ Persei association is expanding. From the expansion rate he managed to estimate the age of the association at about 1 300 000 years. The total life of an association is no more than 10^8 years. Given that there are about 10 000 of them, they must be continually created. It is believed that every one thousand years a new association is created out of the interstellar medium.

1.6 Interstellar Medium

The space between the stars is not empty but is filled by some very diffuse matter called the *interstellar medium*. Its density is a million times less than the density of the "vacuum" we can create in our laboratories using our most efficient vacuum pumps. Even so, due to the enormous volumes over which the interstellar matter extends, its total mass is still very high. Of the total galactic mass in the solar neighbourhood, 63 % belongs to the stars and 37 % is in the form of interstellar matter. Its density is about 0.8 atoms/cm^3 or 1.3×10^{-24} g/cm^3. (The density of the atmosphere on the surface of the Earth is about 10^{19} atoms/cm^3.) The interstellar matter consists of gas and dust. The dust is 100 times less dense than the gas, with an average density of about 10^{-26} g/cm^3. The abundance of the various chemical elements in the interstellar medium is roughly the same as that in the stellar atmospheres. The hydrogen and helium together constitute 96–98 % of the total mass while the hydrogen by itself is 70–80 % of the mass. Since the time of the formation of our Galaxy the interstellar medium has been enriched with metals by a factor 100–1000. (In astronomy by metals we mean all elements heavier than helium.) These metals were formed in the stellar interiors and spread throughout space by supernova explosions. Only 5 % of the interstellar atoms and molecules are ionised.

The interstellar medium also contains clouds where the local density is significantly above the mean. These *interstellar clouds* or *nebulae* are usually found in the galactic plane and in particular along the spiral arms. Their densities vary between 10 and 10^5 atoms/cm^3.

The interstellar gas clouds can be detected either from their emission lines, when they are excited by nearby bright stars, or by absorption lines detected in the spectra of stars that lie behind them. They may also be detected by their emission at radio wavelengths, which consists of lines if the gas is neutral, and is continuous if the gas is ionised.

1.7 Interstellar Clouds

A high percentage of interstellar space is occupied by various types of interstellar clouds distinguished by their appearance:

1.7.1 Bright Nebulae

There are two kinds of bright nebulae, emission and reflection, according to the way in which they radiate. In all cases their radiation is triggered by nearby stars.

i) *Emission Nebulae*. These are concentrations of ionised hydrogen, written symbolically as HII, close to very hot stars (of spectral type O-B1). The ultraviolet radiation from the stars ionises the gas which then recombines and radiates at certain wavelengths in both the optical and radio bands. The dimensions of these nebulae are 0.05–200 pc. An example is the Orion nebula (Fig. 1.8).

ii) *Reflection Nebulae*. If the nearby star is of a later spectral type (e.g. B2-A) its ultraviolet radiation is insufficient to ionise the nebula. Instead, the *interstellar dust* particles within the nebula reflect and diffuse the radiation they receive.

We receive both reflected and emitted radiation from many bright nebulae. Many of them also have dark patches, which are cold, dense concentrations of matter projected onto the bright parts. In some bright nebulae we observe some dark, round clouds called *globules* (Fig. 1.9). Their masses are about 30 $M_\odot$ and their diameters 0.005–1 pc. They are believed to be the progenitors of stars, the sites where stars collapse out of the interstellar gas and dust. The timescale for such a collapse is very small in comparison with the age of the stars, of the order of a few thousand years.

An interesting subgroup of the bright nebulae are the *planetary nebulae,* which owe their name to their appearance when seen through a telescope with small magnification. They are rings or shells of material of diameter 0.1–1 pc. Their density is $10^3 – 10^5$ atoms/cm^3. At their centre we usually find a bright star with a surface temperature of 50000–100000 K. The light we receive from the nebula is the result of emission triggered by the ultraviolet radiation of the central bright star. It is believed that the

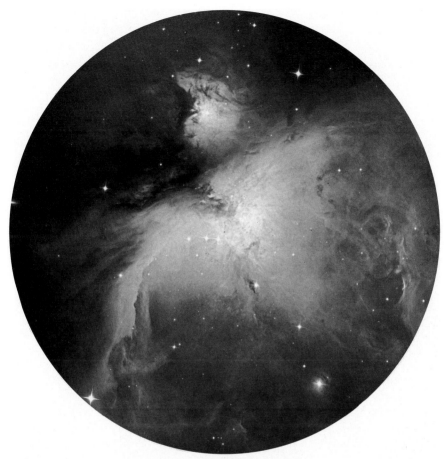

Fig. 1.8. The Orion nebula at a distance of 1600 ly with a diameter of 20 ly. It is illuminated by four hot, blue stars which are inside its gaseous mass (European Southern Observatory)

planetary nebulae are created by an explosive release of matter from the central star.

1.7.2 Dark Nebulae

They appear as extensive dark areas in the sky, having diameters of a few hundred light years. There are no nearby bright stars and the stars behind them appear dimmer and redder than normal, with their light partially polarised (by about 1–7%). *Interstellar absorption, interstellar reddening* and *polarisation* of the stellar light are all due to dust grains in the nebulae, while the gas only absorbs light from the background stars at selective wavelengths, and hence superimposes absorption lines on the spectra of

Fig. 1.9. The Eagle nebula (M 16) in the constellation of Monoceros. It is rich in dark "globules" which are the first stages of star formation (Palomar Observatory photograph)

these stars. Such absorption lines, which are caused by interstellar gas, are called *interstellar lines*. About 10 % of interstellar space is filled with such nebulae, the masses of which vary between one and a hundred solar masses.

1.7.3 Neutral Hydrogen Nebulae

These nebulae, also called HI regions, cannot be seen in the optical band. They are made manifest by the absorption lines they imprint onto the spectra of stars lying behind them. From such evidence we deduce that there is a large number of HI regions.

In 1944 *van de Hulst* predicted that we should be able to observe the interstellar neutral hydrogen at a wavelength of 21 cm, in the radio band. Indeed, a few years later, in 1951, several radio astronomers from the USA, Netherlands and Australia observed the 21 cm-line of neutral hydrogen. Since then many other emission lines due to interstellar molecular hydrogen, oxygen, hydroxyl (OH) etc. ..., have been predicted and observed with radio telescopes. Radio astronomers can deduce from these observations not only the temperature of the gas (about 120 K), and the direction of the HI region but also its distance from us. The gas is not uniformly distributed in the Galaxy but concentrates towards the galactic plane, especially along the spiral arms.

1.8 Radio Waves from Our Galaxy

Radio waves from space were discovered by *Jansky* during some experiments in 1931 at the Bell Telephone Laboratories. The first parabolic radio telescope was built in 1936 by *Reber* who also constructed the first radio-maps of our Galaxy. The construction of really large radio telescopes started after the Second World War, in Britain, Netherlands, Australia and later the USA, Germany and the USSR. Thus radioastronomy was born. The Earth's atmosphere, including the clouds, is transparent at wavelengths greater than 1 cm and less than 30 m. This range of wavelengths is known as the *radio window,* as opposed to the *optical window* which contains the range of wavelengths to which the human eye is sensitive. Radio telescopes can observe from 1 mm to about 100 m. The interstellar nebulae are almost transparent to radio waves, thus allowing us to receive signals from very large distances, much further than we can see with the optical telescopes.

1.8.1 Radio Sources

There are some regions in our Galaxy from which we receive very intense radiation at radio wavelengths. Some of the strongest radio sources are the nebulae which resulted from supernova explosions, such as the radio source Cassiopeia A, the source in the constellation of Taurus which has been identified with the Crab nebula (Fig. 1.10), etc. Some radio sources are identified with galaxies, and we shall discuss them later (Sects. 2.13, 2.14).

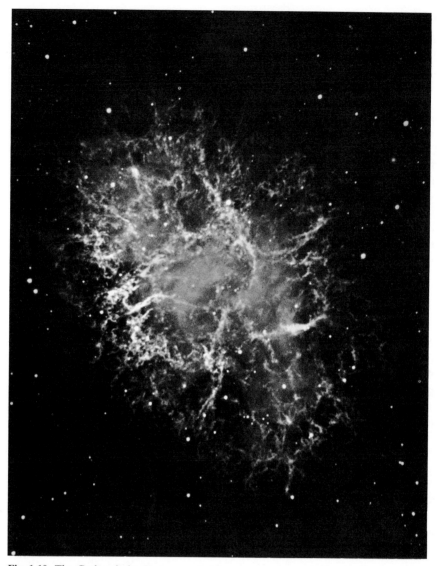

Fig. 1.10. The Crab nebula. It was created by a supernova explosion in 1054 A.D. (Lick Observatory)

1.8.2 The 21cm-Line and the Distribution of Hydrogen in the Galaxy

The radiation we receive from the interstellar medium in the radioband is either at discrete wavelengths or continuous. The most important emission line we observe is the 21cm-line of neutral hydrogen. This line is due to the hyperfine splitting of the ground state of the hydrogen atom. Due to the

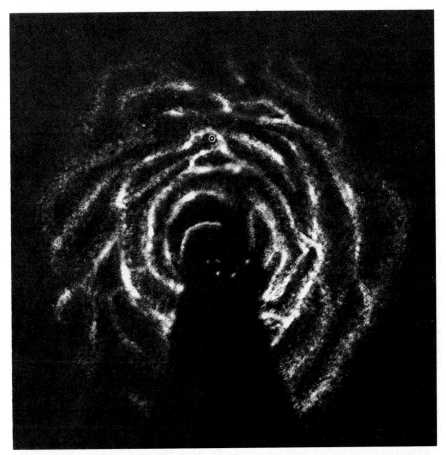

Fig. 1.11. The appearance of the Galaxy as observed with radio telescopes from the Earth at the wavelength of 21 cm. The regions towards the centre cannot be observed accurately and are left empty (courtesy of Dr. G. Westerhout)

spin of the electron being either parallel or antiparallel to that of the nucleus, the ground state of the hydrogen atom is split into two components. The energy difference between these two components is very small. When the electron drops from the parallel position to the antiparallel one, the photon released has a very small frequency ($v = 1420.6$ MHz) as calculated from the formula $\Delta E = hv$, where h is Planck's constant, v is the frequency of the photon emitted and ΔE is the difference between the two energy levels. This transition happens only once in 11×10^6 years per one hydrogen atom, but since the quantity of interstellar hydrogen is very large, we are able to observe the 21cm-line very clearly. This line is used to find the distribution of neutral hydrogen in the Galaxy. A map of our Galaxy in the 21cm-line is given in Fig. 1.11.

1.8.3 Recombination Lines

Recombination is the process during which an ionised atom combines with an electron to become neutral. Usually the electron is captured into a high energy state, from which it gradually cascades down to the ground state. During this transition from one level to the next it emits radiation at various wavelengths. The spectral lines produced in this way are called *recombination lines*. Since the higher energy levels are very close to each other, the energy difference between them is small and thus the wavelengths of the recombination lines are large, reaching into the radio band. We have observed several recombination lines in the spectra of bright nebulae, due to hydrogen and helium atoms. Several of them appear in the optical and infrared bands, but most of them are of centimetre wavelength. These latter lines are those used to estimate the distribution and distance of the interstellar matter.

1.8.4 Microwave Radiation from Molecules

Observations at millimetres or a few centimetres wavelength (microwaves) have shown the existence of dense, cold clouds containing composite molecules of interstellar matter. More than 50 molecules have been discovered in interstellar space. Some of them are inorganic like water, ammonia, hydrogen sulphide etc., and some are organic such as formaldehyde, formic acid, ethanol etc. Molecular clouds are the sites of the formation of young stars. The total mass of the molecular hydrogen is estimated to be as great as that of the atomic hydrogen in the Galaxy.

1.8.5 Continuous Radio Emission

We receive continuous radiation mainly from the spiral arms of our Galaxy. Similar radiation is received from other galaxies also (Sect. 2.13). This continuous radio emission can be either *thermal* or *non-thermal*.

In the HII regions, radio waves are produced by collisions (encounters) between electrons and protons. The electrons are accelerated and deflected from their initial paths but are not captured. The radiation produced is continuous, since the electron, unlike when it is bound, does not jump between two discrete energy levels. Such collisions, during which the electron remains free, are called *free-free transitions*. The radiation produced in this way is called *thermal radiation* since it depends mainly upon the thermal motions of the electrons in the region.

On the other hand, *non-thermal radiation* arises due to the motion of very energetic electrons inside a magnetic field. These electrons, called relativistic electrons, move with speeds approaching that of light. The radiation produced in this way is called *synchrotron radiation* because it was observed for the first time in a synchrotron accelerator.

In Table 1.1 we summarise the various forms of interstellar matter.

Table 1.1. Observable forms of interstellar medium

Region of the spectrum	Gas	Dust
Optical emission	HII regions	Reflection nebulae
Optical absorption	Interstellar clouds (absorption lines in stellar spectra)	Dark nebulae, absorption, reddening and polarisation of light from stars behind the nebulae
Radiowaves	HI 21cm-line Recombination lines Continuum radiation	
Microwaves	Molecular lines	

1.9 x-Ray, γ-Ray, Ultraviolet and Infrared Astronomy

In recent years the development of rockets, artificial satellites and space-ships has enabled us to examine astronomical objects in the x-ray, γ-ray, ultraviolet and infrared wavebands. These modern branches of astronomy explore the Universe from above the Earth's atmosphere and have, to date, produced a wealth of observational data.

The x-ray satellite Ariel 5 has produced a map containing about 300 x-ray sources (Fig. 1.12). Most of them are in our Galaxy, but there are some extragalactic sources too. The most intense source of x-rays lies in the constellation of Scorpio, known as Sco X-1. It is a double star with one member likely to be a black hole. The galactic x-ray sources detected by

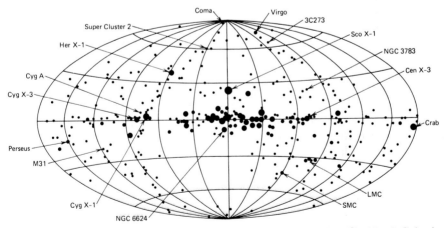

Fig. 1.12. The x-ray sky. Stronger sources are marked with larger dots. The North Galactic Pole is at the top of the map

the Uhuru satellite have been identified as supernova remnants (e.g. the Tycho supernova, the Crab nebula etc.), as tightly bound binary stars, or as globular clusters. The Einstein satellite, which was launched in 1978, has produced plenty of accurate observational data in the x-ray region.

γ-ray observations have been used to investigate the Sun, the planets, the interstellar medium and generally our Galaxy, as well as clusters of galaxies. The diffuse γ-radiation we receive from the galactic plane is believed to be produced by the interaction of cosmic rays with the interstellar medium, especially atomic and molecular hydrogen.

With the help of satellites we can also observe in the ultraviolet, which corresponds to wavelengths between the optical and the x-rays, as well as in the infrared, between the optical and radio windows. This radiation does not reach the surface of the Earth as it is absorbed by the Earth's atmosphere. Many new spectral lines were found in the ultraviolet. On the other hand, many objects with strong infrared emission have been observed. One of them is in the Orion nebula, which may be a star in the first stages of its life. The strongest observed source in the infrared is at the centre of our Galaxy, in the region of the constellation of Sagittarius.

1.10 Pulsars (Neutron Stars)

In 1967, *Hewish* (Nobel prize 1974) and *S.J. Bell* discovered some pulsating radio sources. They are variable stars with a very short period ranging from 0.0015–3 seconds. These periods are so unusually short, that when the first signals from pulsars were received, some astronomers thought they were broadcasted by extraterrestrial civilisations.

According to the currently accepted theory, the density inside a pulsar is of the order of $10^{11}–10^{14}$ g/cm^3. At such high densities the electrons and protons are united into neutrons. This is why pulsars are also known as *neutron stars*.

Pulsars are roughly as massive as the Sun, but with diameters of only 10–20 km. They have very strong magnetic fields (10^{12} Gauss) which rotate with the star. Two pulsars have been photographed close to the centres of supernova remnants. The NPO 532 pulsar is in the Crab nebula (Fig. 1.10) and was observed and photographed in the optical in 1969.

A pulsar is the remnant of a star after a supernova explosion. During the explosion a large amount of energy is suddenly released expelling the outer layers of the star. The main body of the star contracts. During this process the star loses about one solar mass of material and becomes a hundred thousand times brighter than the Sun.

Observations over a long period of time have shown that the periods of pulsars systematically increase with time. The stars gradually lose some of their energy and angular momentum and rotate more slowly. It is believed that the surface of the pulsar is solid (crystalline) while their nucleus is superfluid. Sudden reductions in the period are occasionally observed and these may be due to changes on the surface of the star, like mini-earthquakes. Pulsars also produce cosmic rays.

It seems that pulsars are the most important physics "laboratories" which exist in the Universe apart, of course, from the Big Bang itself.

1.11 Magnetic Fields and Cosmic Rays

1.11.1 Magnetic Fields

The light we receive from the stars is often polarised. The greater the interstellar absorption by intervening dust the greater is the degree of *polarisation*. This polarisation is due to *interstellar magnetic fields*. We obtain further information about the magnetic fields in the Galaxy from radio observations. Radiation from some galactic radio sources passes through ionised gas permeated by a magnetic field. This polarises the radiation in different directions at different wavelengths. This is known as Faraday rotation. The presence of magnetic fields can also be detected from the Zeeman splitting of the 21 cm-line, or from synchrotron radiation from electrons moving inside magnetic fields.

The magnetic field lines in our Galaxy are aligned along the spiral arms. The magnetic field strength, either measured from interstellar absorption or from the synchrotron radiation, turns out to be about 2×10^{-6} Gauss. (The Earth's magnetic field is 0.5 Gauss.) The energy density of the magnetic field is approximately 10^{-12} erg/cm^3.

1.11.2 Cosmic Rays

Cosmic rays reach the Earth from all directions. They consist of high-energy particles, mostly protons and electrons but also nuclei of helium and other heavy elements, travelling at very high velocities. On reaching the Earth's atmosphere they produce a plethora of new particles upon collision with atmospheric atoms and molecules, which reach the surface of the Earth. Such products of spallation reactions are called "secondary" cosmic radiation.

The sources of cosmic radiation comprise the Sun, our Galaxy and even extragalactic objects. The cosmic rays are therefore samples from distant parts of the Universe. Cosmic radiation from the Sun has very low energy (up to 10^{10} eV = 10 BeV), while higher-energy radiation comes from the Galaxy (up to 10^7 BeV), and is constrained to follow the galactic magnetic field lines. Sources of cosmic rays are usually sites where there has been some violent activity such as the formation of a nova, supernova or pulsar. Because of their high energy, cosmic rays heat and ionise the

Table 1.2. Energy density in our region of the Galaxy

Cosmic radiation	1×10^{-12} erg/cm³
Total stellar light	0.7×10^{-12} erg/cm³
Kinetic energy of the interstellar medium	0.4×10^{-12} erg/cm³
Energy of the galactic magnetic field	1×10^{-12} erg/cm³

interstellar gas. The total energy density of cosmic rays is 10^{-12} erg/cm³. This is about the same as the total energy density of starlight, i.e. the energy density of light from all the stars of our Galaxy at all wavelenghts. The kinetic energy density of the interstellar gas is also of the same order of magnitude (see Table 1.2).

1.12 Stellar Populations

Baade divided the stars of our Galaxy into two populations. Later, more subdivisions were made. The most important stellar populations are the following:

1) *Population I*. These are very young stars with a chemical composition rich in metals (2–4% of their total mass). They are abundant along the spiral arms, and have nearly circular orbits about the galactic centre. The stars in open clusters and OB associations are population I stars.

2) *Disc Population*. This is a population which is intermediate between populations I and II (see below). Most of the stars close to the galactic plane belong to this population. They are distributed in a disc around the galactic centre, and their motions are almost circular.

3) *Population II*. They have weak metal lines, having only 10% of the metal abundance of population I stars (0.3% of their total mass). The RR Lyrae stars and the stars in globular clusters are population II stars. Stars of the galactic halo also belong to this population (Fig. 1.2). Their orbits about the galactic centre are elongated and their ages are at least 10^{10} years.

In recent years a hypothetical population III has been introduced to designate stars that were formed before the formation of galaxies (Sect. 8.3).

1.13 The Spiral Shape of Our Galaxy

The main characteristic of our Galaxy is its spiral structure. Although it is difficult to determine such structure in our own Galaxy, in recent years the distribution of matter to large distances has been probed and the spiral structure established. The spiral arms of our Galaxy were discovered in 1951 from observations in the optical by *Morgan, Nassau* and their collaborators. They studied the distribution of HII regions and of O stars up to a distance of a few kpc and discovered the Perseus, Orion and Sagittarius spiral arms. The names were chosen from the constellations where the bright O-B 2 stars of the arms were projected. Their results were confirmed by more recent observations. The distance between the spiral arms is approximately 2 kpc and their thickness 500 pc.

The most important observations of the spiral arms have been carried out by radio telescopes. The observations in the 21 cm-line allowed us to calculate the distances of the HI regions and from the radiation maxima the positions of the spiral arms were found. We now know that the arms do not have constant density. A certain condensation of the Orion arm close to the Sun is particularly interesting and is called the Local System, or Gould's Belt. The spiral arms of our Galaxy are not entirely flat, because the disc is warped. That is, part of the disc curves up towards the North Galactic Pole, while the diametrically opposite part curves towards the South Galactic Pole. Figure 1.11 shows the general picture of the distribution of hydrogen in the galactic plane. We see that the distribution is far from smooth. This map was made by Dutch and Australian radioastronomers. We can further study the spiral structure from the interstellar absorption lines which are due to atoms or simple molecules.

Observations of the central parts of our Galaxy have shown that there is a gaseous disc of neutral hydrogen at the centre. Its mass is of the order of 10^8 solar masses and its radius 600 pc. It rotates with velocity 200 km/sec at radius 100 pc. Its rotation velocity increases outwards and reaches up to 250 km/sec. The spiral arms of the Galaxy emanate from two diametrically opposite points of the nucleus (Fig. 1.3).

1.14 Kinematics of the Galaxy

1.14.1 Motion of the Sun

The apparent motions of the stars in the solar neighbourhood are partly due to the peculiar motions of the stars themselves and partly due to the motion of our Sun. If we consider the average velocity of the stars in the solar neighbourhood, the peculiar motions average out and we are left with a velocity equal to that of the Sun but with opposite sign. We find that the Sun moves with a velocity $V_\odot = 20$ km/sec towards a point with coordinates $\alpha = 18^h$ and $\delta = 30°$, called the *apex*. Due to this motion the neigbouring stars appear to move towards the *antapex*, i.e. directly away from the apex on the celestial sphere.

1.14.2 Distribution of Stellar Velocities

The peculiar velocities of the stars are not randomly distributed. They show some preference for motions towards two diametrically opposite points called *vertices*. The line which connects them lies on the galactic plane and passes close to the galactic centre. The implication of this is that the stars move preferably towards or away from the galactic centre. This phenomenon can be understood in terms of the ellipsoidal distribution of velocities suggested by *K. Schwarzschild*. Consider a parallel displacement of the peculiar velocity vectors of the nearby stars so that they have the same origin. The tips of these vectors will then mark out a distribution of points around the origin. The isodensity contours of these points are similar ellipsoids with their major axes pointing towards the centre of the Galaxy.

This regularity is observed for stars with peculiar velocities of the order of 20 km/sec. The distribution of stars with much higher velocities is rather asymmetric. Almost all the high velocity stars move in a direction opposite to the Sun's motion with respect to the galactic centre. As found by *B. Lindblad* (1926), this observed asymmetry can be understood if we realise that the velocities of the stars with respect to the galactic centre are the geometric sum of their observed velocities plus the velocity of the Sun with respect to the galactic centre. Therefore the high velocity stars we observe have small rotational velocities. These are the population II stars which move on elongated orbits passing close to the galactic centre.

1.14.3 Rotation of the Galaxy

Oort was the first who studied the rotation of the Galaxy systematically and came to the conclusion that the Galaxy does not rotate like a solid body. A solid body rotates with constant angular velocity Ω, while the angular velocity of our Galaxy is a function of the distance R from the galactic centre (i.e. $\Omega = \Omega(R)$). We say that our Galaxy has *differential rotation*. The stars which are closer to the galactic centre rotate faster than those which are further out. This is similar to what happens in the solar system where the outer planets rotate more slowly than the inner ones (Kepler's third law).

The population I stars are in almost circular orbits about the galactic centre. Our Sun, for example, is in an almost circular orbit with velocity 220 km/sec. The deviations from circular velocity are small.

By studying the motions of stars in our Galaxy we can learn the following:

– If we know the law $\Omega = \Omega(R)$ we can estimate the distance of the interstellar medium using observations in the 21 cm-line, or the recombination lines.

−   From the rotational velocities one can calculate the total mass of the Galaxy as well as its mass distribution.

Assume that the total mass M of the Galaxy is concentrated near its centre, and that the stars in the solar neighbourhood move in circular orbits. Equating the centrifugal force $\Omega^2 R$ to the gravitational attraction from the Galaxy GM/R^2 we obtain

$$\Omega^2 = \frac{GM}{R^3} \qquad (1.1)$$

where Ω is the angular velocity at radius R and G $(= 6.7 \times 10^{-8}$ cm^3 sec^{-2} g$^{-1})$ is the constant of gravity.

In order to find the *distribution of matter in the Galaxy* from the observed velocity curve, we construct a model of the Galaxy, compute the motions of the stars in it, and compare them with observations.

1.15 Dynamics of the Galaxy

We divide the force acting on a star into two parts: (1) The force due to the smoothed distribution of matter in the Galaxy as a whole. (2) The force due to individual neighbouring stars. A star's orbit is fixed mainly by the smoothed out force while the second component acts as a perturbation to this motion. The *relaxation time* is the time required for a star to be deflected through 90° from its initial orbit by the perturbation of neighbouring stars. The relaxation time for open clusters is of order 5×10^7 years. Due to the influence of nearby stars certain cluster members may acquire velocities in excess of the escape velocity and thus leave the cluster. If T_0 is the relaxation time of a cluster, it can be shown that 90% of the cluster members will have escaped after 42 T_0. Therefore the total life of a cluster is 5×10^9 years. The Pleiades will have lost half of its members in 2×10^9 years. The globular clusters on the other hand may survive for many billions of years whereas the associations live for only a few million years.

The evolution of our Galaxy is a more difficult problem. Our Galaxy was formed 10–20 billion years ago, just after the Big Bang (Sect. 6.9). The first stars and the globular clusters formed at roughly the same time in a collapsing cloud of gas. Most of the stars of the Galaxy were formed during the first few billion years of the Galaxy's lifetime, but new stars are formed continually in the interstellar clouds.

Our Galaxy will eventually disintegrate. It is estimated that in about 10^{14} years all stellar systems, including clusters and galaxies will have dispersed.

1.16 A General Picture of the Galaxy

We may say that our Galaxy consists of a bulge, a disc and a halo. The bulge is essentially a gigantic oblate cluster. Its diameter in the galactic plane is about 2 kpc while perpendicular to it it measures 1.5 kpc. Its density increases dramatically towards the centre. It is estimated that at 0.1 pc from the centre the density is 100 million times greater than the density in the solar neighbourhood. If the stellar density were so great locally, then the Earth would be illuminated as if from 100 full moons. This gigantic cluster includes 10 billion stars in total. Nevertheless, even in this region of the Galaxy collisions between stars are very unlikely. Apart from the stars there is gas, mainly hydrogen. In the outer parts the gas is in molecular form. Closer to the centre most molecules are broken down into atoms and still closer the atoms are ionised. The total mass of the gas is of the order of 100 million solar masses, while the dust is 100 times less.

The disc stars are more or less smoothly distributed in the galactic plane at large distances from the galactic centre, about which they rotate in almost circular orbits. The disc, including interstellar matter and stars, contains 1/5 of the total mass of the Galaxy. It has a diameter of 30 kpc and a thickness of at least 2 kpc at the centre, becoming thinner in the outer parts (Fig. 1.2). The spiral arms start near the galactic centre, and reach out to 15 kpc. Their thickness is about 0.5 kpc and they consist of young stars, stellar associations, open clusters as well as interstellar gas and dust. The mean density in the arms is about 10 atoms/cm^3, much higher than the density in the regions between the arms which is only 0.8 atoms/cm^3. In other words the spiral arms are the densest parts of the galactic disc.

Table 1.3. The Galaxy

Diameter of the Galaxy	30 kpc
Diameter of the halo	50 kpc
Thickness of the disc	2 kpc
Thickness of the gas close to the galactic plane	250 pc
Mass of the Galaxy	$10^{12} M_\odot$
Mass of the galactic disc	$2 \times 10^{11} M_\odot$
Average density	$0.1 M_\odot/pc^3 = 7 \times 10^{-24} g/cm^3$
Density close to the centre	$(r = 1 \text{ pc})\ 400\,000\ M_\odot/pc^3$
	$(r = 10 \text{ pc})\ 7000\ M_\odot/pc^3$
	$(r = 100 \text{ pc})\ 100\ M_\odot/pc^3$
Distance of Sun from galactic centre	8.5 kpc
Distance of Sun from galactic plane	14 pc
Rotation velocity at position of Sun	220 km/sec
Time for one rotation of Sun about the galactic centre	240×10^6 years
Escape velocity from galactic centre	700 km/sec
Escape velocity from solar neighbourhood	350 km/sec

The Galaxy rotates in the clockwise direction when seen by an observer on the North Galactic Pole. The whole system behaves like a winding up spiral. For this reason we say that the Galaxy has "trailing" spiral arms.

The bulge and the disc are embedded inside a slightly non-spherical nebula called the halo (Fig. 1.2). This includes the globular clusters, a large number of population II stars, among them the RR Lyrae variable stars, as well as diffuse interstellar gas. The total mass of the halo is $10^{12} M_\odot$, much larger than the mass of the disc, which is about $2 \times 10^{11} M_\odot$. The orbits of the halo objects are different from those of population I and disc stars. In general they follow highly inclined, elongated orbits about the galactic centre, hardly participating in the rotation of the Galaxy. Table 1.3 summarises certain basic characteristics of our Galaxy.

2. Galaxies and Clusters of Galaxies

2.1. Classification of Galaxies

Numerous galaxies similar to our own can be observed at distances far beyond the boundaries of our Galaxy. Sometimes they are referred to as "extragalactic nebulae" because they look like faint nebulae when seen through small telescopes. Consequently they were initially given names such as the "Magellanic Clouds", the "Andromeda nebula" and so forth.

Observational astronomy advanced hand in hand with the technology to build increasingly powerful telescopes. In 1610, *Galileo* became the first person to observe the sky with a telescope. It was not until the 18th century, however, that the nebulae could be distinguished from the stars of our Galaxy. The philosopher *Kant* concerned himself with astronomical problems early in his life, and in his classic work "Universal Natural History and Theory of the Heavens" (1755) he wrote: "The analogy (of the nebulae) with the system in which we find ourselves ..., is in perfect agreement with the concept that these elliptical objects are just (island) Universes, in other words, Milky Ways". At the end of the 18th century, observations made by *W. Herschel* and his son, revealed hundreds of such *islands of the Universe*. Our own Galaxy was considered to be a large continent. Later on, still larger telescopes showed that our Galaxy is very similar to the other galaxies in the "ocean of the Universe".

The first systematic study of the galaxies was done with the 2.5 m-telescope at Mount Wilson, by *Hubble* (1889–1953). The investigation continued with the aid of Schmidt telescopes and the large Lick and Palomar telescopes.

Hubble classified galaxies according to their morphology into four basic categories: *Ellipticals* (E), *spirals* (or normal spirals, S), *barred* (or barred spirals, SB) and *irregulars* (Ir). Later on a separate category of *lenticulars* (S0) was appended as an intermediate type between ellipticals and spirals. Lenticular galaxies generally look like spirals with large bulges but without spiral arms (Fig. 2.1). Elliptical galaxies owe their name to their shape. They are divided into eight categories (E0, E1, ..., E7)[1].

---

[1] The number next to the letter E is ten times the oblateness of the elliptical image of the galaxy, defined as the ratio $(a - b)/a$, where a and b are the major and minor axes of the ellipse respectively.

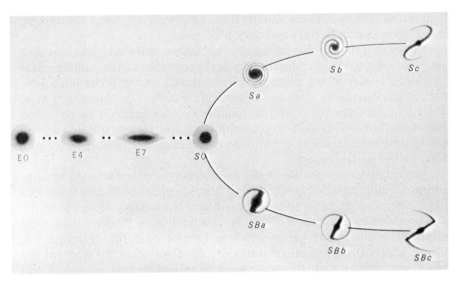

Fig. 2.1. The classification of the galaxies according to *Hubble*

They consist of population II stars and contain very little gas and dust. Their diameters vary between 1 and 150 kpc. We generally distinguish between dwarf and giant ellipticals. Giant ellipticals have absolute magnitudes brighter than -21, while dwarf ellipticals are fainter than magnitude -14. There are also some elliptical galaxies of intermediate size.

Ellipticals used to be considered as oblate spheroids flattened by rotation about their short axis. In recent years, however, observations have shown that the rotation of giant ellipticals is too small to be consistent with this picture. Thus giant ellipticals are probably triaxial ellipsoids, more elongated along a particular axis.

The normal spiral galaxies are of types Sa, Sb or Sc. All of them have spiral arms which start tangentially from two opposite points of the nucleus. The various types are distinguished by the appearance of their spiral arms, which are progressively more open from types Sa, Sb through to Sc. These three types have their counterparts among the barred spiral galaxies: SBa, SBb and SBc. In barred spiral galaxies the spiral arms start perpendicularly from the ends of the bar. The spiral arms in SBa galaxies are so tightly wound that they appear to close in a ring around the bar, looking like the greek letter Θ. On the other hand, SBc galaxies look very much like a letter S. There are also some dwarf spiral galaxies, but these are not very common. Along the Hubble sequence the content of young objects and interstellar matter increases, from ellipticals to spirals. Sc galaxies contain more population I objects than Sb and Sa galaxies.

Finally, there are two kinds of irregular galaxies: population I and population II. Population I irregulars contain a lot of gas and dust.

Population II irregulars are usually small companion galaxies, containing mainly population II stars and very little gas.

We see that the Hubble sequence is not only a morphological sequence but also a sequence of changing physical properties of the galaxies. Star formation is still taking place in spiral galaxies and in some irregulars, while it has essentially stopped in ellipticals. Further, the amount of mass needed to produce the observed luminosity of a galaxy (the so called mass-to-light ratio) is high among ellipticals, less among spirals and still less among irregulars. The implication is that elliptical galaxies have more mass locked in faint stars.

According to *van den Bergh* the distribution of galaxies in the various groups is: ellipticals 30%; spirals and barred spirals 60% and irregulars 10%. The proportion of spirals quoted here is probably higher than the true one, due to a selection effect. Spirals tend to be brighter than most ellipticals, while dwarf galaxies, which cannot be easily observed, are almost invariably elliptical or irregular. So if we take into account the dwarf galaxies as well, the percentage of ellipticals and irregulars will increase.

A characteristic property of a galaxy is the degree of its rotation. Theoretical studies suggest that the various shapes of galaxies are due to the various values of some basic parameters, i.e. the initial mass and rotation of the cloud of matter which evolved into the galaxy.

Three other classification schemes have been suggested in recent years, by *de Vaucouleurs, Morgan* and *van den Bergh*. Even so, the Hubble classification is still widely used.

2.2 Observations of Galaxies

Many powerful telescopes, capable of observing distant galaxies have come into use in recent years. Many factors besides the use of larger mirrors have contributed to this: The quality of the materials used and the technique by which the mirror's surface is ground have been improved; new combinations of lenses and mirrors have been invented; special photographic techniques have been developed and more electronic systems have been employed for greater accuracy. Some new emulsifiers have been found which have immensely improved the developing of photographic plates. In order to photograph large regions of the sky one uses special telescopes, the so-called Schmidt telescopes. Such a telescope is at the Mount Palomar Observatory and has a diameter of 180 cm. Stars 200 000 times fainter than those visible to the naked eye can be photographed with this telescope. It has been used in compiling the "Sky Survey". This survey consists of two series of photographs covering the sky from the North Pole down to declination $- 30°$ (which is the southernmost declination visible from the Palomar Observatory). One series has been taken in the blue waveband, and the other in the red. Objects as faint as 21st magnitude are visible on the blue plates while the red plates show objects down to 20th magnitude. A similar telescope, built ten years ago in Siding Springs (Australia), can photograph objects down to 23rd magnitude. This makes the Australian telescope the most powerful of its kind in the world since it can detect objects 2.5 times further away than the Palomar telescope.

With the aid of modern telescopes we can distinguish individual objects in galaxies which are up to a few million light years away. Thus we observe stars, including variable stars, novae, supernovae, planetary nebulae, associations, open and globular clusters and the interstellar medium (bright nebulae, dust, neutral and ionised hydrogen). In general we find all the objects we see in our own Galaxy. It is particularly useful to be able to observe novae and supernovae in external galaxies as we can calculate distances from them. Supernovae can reach luminosities comparable to those of their parent galaxies, and so can be observed in very distant galaxies.

2.3 Magellanic Clouds

The Magellanic Clouds are two small galaxies which are considered as companions, or satellites, of our own Galaxy (Figs. 2.2a, b). From the southern hemisphere they can be seen by the naked eye, since they are almost as bright as the Milky Way. Because they are the closest galaxies to our own they have been studied extensively for many years. The Large Magellanic Cloud (LMC) is about 50 kpc away from us, while the Small Magellanic Cloud (SMC) is at a distance of 55 kpc. In comparison with other extragalactic objects, however, they are very close to us. Since we know their distances we can estimate their diameters which turn out to be 6.5 and 2.9 kpc respectively. Given that the diameter of our Galaxy is 30 kpc, we conclude that they are indeed small galaxies. Even so, because of their proximity they have caused a noticeable effect on our galactic plane, which is warped.

At first sight both Magellanic Clouds appear to be irregular. Unlike our own Galaxy, they do not have dense bulges or spiral arms. A more detailed study shows, however, that the LMC has most of the characteristics of a barred spiral.

Both Clouds contain a large quantity of interstellar matter. One tenth of the mass of the LMC and as much as a third of the mass of the SMC consist of gas and dust in the form of bright and dark nebulae. The diameters of both of them are larger when viewed at radio wavelengths. From radio observations in the 21 cm-line we know that both contain neutral hydrogen.

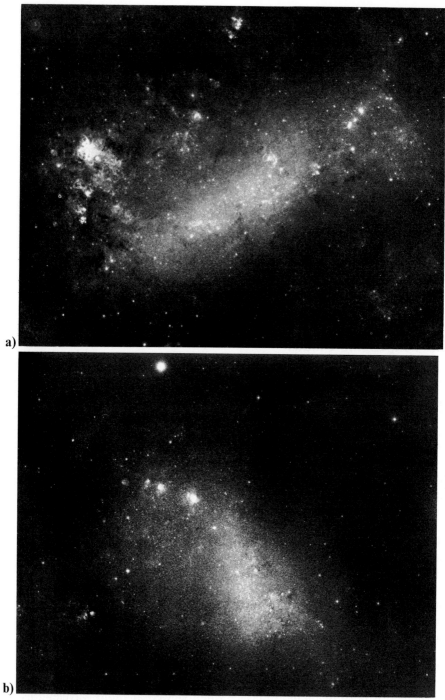

Fig. 2.2. (a) The Large Magellanic Cloud is at a distance of approximately 50 kpc and has a diameter of 6.5 kpc. There is a bar in the middle of the Cloud (European Southern Observatory). **(b)** The Small Magellanic Cloud is at a distance of 55 kpc and has a diameter of 2.9 kpc (European Southern Observatory)

The radio observations also give us information about some of the kinematic properties of the two Clouds. They both rotate about their common centre of gravity with a period which appears to be equal to the periods with which they spin about their own axes. In this way they always show the same "face" to each other. That is, they rotate as a dumbell connected by an invisible rod.

The Uhuru satellite has detected four x-ray sources in the LMC and one in the SMC. Two of these sources are very strong, being 10 times stronger than the Crab Nebula. We also receive diffuse x-ray radiation from both Clouds. The total power emitted by the SMC is one tenth that emitted by our own Galaxy.

Studying the Magellanic Clouds is not only intrinsically interesting, but also helps us to understand better the structure of our own Galaxy. As we are outside observers of these galaxies, and yet are nearby, we can more easily form a picture of the distribution of stars, clusters and nebulae in them, than in our own Galaxy. An example of the useful information we can obtain from studying these galaxies is the work done by *Leavitt* (1912) on the cepheid variables of the SMC. Her research led to the period-luminosity relation which was the key to calculating the distances to other galaxies.

A lot of research has been done on the clusters and HII regions of the Clouds. In the LMC alone we know of more than 1600 star clusters and 500 HII regions.

Spectroscopic techniques can be used to specify the compositions of the stars, and the interstellar matter of the Magellanic Clouds. A comparison with the corresponding objects in our own Galaxy can give valuable information concerning the origin and evolution of galaxies. This is why we cannot overemphasise the importance of research on the Magellanic Clouds, particularly now that large astronomical instruments have been built in the southern hemisphere.

2.4 The Andromeda Galaxy

This galaxy, also known as M 31, is similar to our own in size, mass, the particular objects it contains, its rotation law and its radio emission. It can be seen with the naked eye as a fuzzy faint patch in the constellation of Andromeda. With a small telescope it looks like a small ellipse of dimensions $1.5° \times 1°$ and looks particularly impressive through a large telescope (Figs. 2.3 a, b). It is at a distance of about 690 kpc and has a diameter of 35 kpc. Its nucleus is like a large star cluster 100 times more massive than an average globular cluster (it contains some 10^7 solar masses). Gas emitted from its nucleus has a total energy of 10^{37} erg/sec, similar to the energy of the gas coming out of the nucleus of our own Galaxy. The bulge is

Fig. 2.3a. The Andromeda galaxy (M 31) is at a distance of 690 kpc. Nearby are two small satellite galaxies (Karl-Schwarzschild-Observatorium, Tautenburg/GDR)

surrounded by a disc with spiral arms, which are rich in stellar associations and HII regions. Also, dozens of x-ray sources have been detected in this galaxy. The disc is surrounded by a halo 100 kpc in diameter, containing more than 350 globular clusters, just like our own Galaxy. The observed mass of the disc is estimated to be $3 \times 10^{11}\ M_\odot$, again comparable to the disc mass of our own Galaxy.

Fig. 2.3 b. A part of the Andromeda galaxy. The galaxy is resolved into a plethora of stars

The Andromeda galaxy is particularly interesting since it is a rich source of valuable information. For example, we can observe its bulge directly in the optical, something which is not possible for our own Galaxy; we can photograph its nucleus, which is also visible in the infrared, and calculate its stellar density. Detailed analysis of the stellar spectra, the gaseous nebulae and the radio emission define accurately the differential rotation of the galaxy. A star at a distance of 10 kpc from its centre takes 250 million years to complete a rotation, a length of time comparable to that required by the Sun to rotate about the galactic centre.

Just like our Galaxy, Andromeda has its own satellite galaxies as well. The most important of them are the ellipticals NGC 205 (Fig. 2.7) and M 32 with diameters of 1.5 and 0.7 kpc respectively. These are the first elliptical galaxies in which individual stars have been observed, by *Baade*. Observations have revealed that the Andromeda galaxy is moving towards us with a velocity of 270 km/sec.

2.5 Spiral Galaxies

Most of the galaxies in the Universe have spiral structure. The spiral arms are usually symmetric, in the sense that one spiral arm would coincide with

Fig. 2.4. The spiral galaxy NGC 5194 (M 51) with its companion NGC 5995 in the constellation of Canes Venatici. The picture of the same galaxy taken at radio wavelengths is superimposed upon the photograph

the other if we rotated it through 180°. There are, of course, various irregularities (Fig. 2.4).

Most of the objects in the spiral arms are young stars and interstellar clouds. The interstellar matter is concentrated near the plane of symmetry of the galaxy (Fig. 2.5). The majority of stars we observe along the spiral

Fig. 2.5. The spiral galaxy NGC 4594 (M 104) in the constellation of Virgo, seen edge on. The dark lane which divides it into two is dust lying in the galactic plane. Around it there are globular clusters seen as bright condensations (European Southern Observatory)

arms are bright giants or supergiants of high temperature, emitting intense ultraviolet radiation. They ionise the gas around them and so create HII regions. When the ionised atoms recombine they emit strong radiation at certain characteristic wavelengths, especially in the $H\alpha$ line of hydrogen, which is in the red. HII regions are particularly prominent in spiral galaxies, our own included. They look like beads along the spiral arms (Fig. 2.4). There is also a lot of neutral hydrogen along the spiral arms, which emits in the 21 cm-wavelength. We also receive continuous radio emission and "synchrotron" radiation due to particles travelling along magnetic field lines. Finally we observe OB associations along the spiral arms.

The understanding of the spiral structure in galaxies has been one of the major astronomical advances in recent years. The most generally accepted theory today argues that the spiral arms are "density waves". That is, the spiral arms are not always made up of the same stars; instead, stars move slowly across them. As the stars pass through the spiral arms their transverse velocity reduces and so they contribute to the local enhancement of density there. The spiral arms are, therefore, the loci of density maxima in the galaxy, but there are many stars outside the spiral arms where the density is lower.

These *density waves* appear very clearly in "numerical experiments" done on computers. These experiments entail the computation of the motion of a large number of particles, sometimes hundreds of thousands, under their mutual gravitational attraction. Each particle may represent a star or an interstellar cloud. At regular time intervals the positions of the particles are displayed pictorially and photographed. A sequence of such frames run consecutively makes a movie presenting the evolution of the model galaxy. Such movies show that an initially irregular concentration of particles results in the formation of two symmetric spiral arms. Detailed observations of the motions of the individual particles confirms the suggestion that the spiral arms are simply the places where the stars move most slowly. Although the spiral arms are not material arms but density waves, they are remarkably stable. They rotate slowly in comparison with the particles which make them up. The sense of rotation is such that the spiral arms go round the centre leaving behind their loose ends. For this reason they are called trailing.

The density wave theory explains a paradox that has often been discussed in recent years: how spiral arms can persist in galaxies with differential rotation. As explained in Sect. 1.14.3 stars closer to the centre of the galaxy rotate faster than those further out. The same phenomenon is observed in the solar system where the outer planets rotate more slowly than the inner ones. The mathematical expression of this phenomenon is *Kepler*'s third law which states that the ratio of the squares of the orbital periods of two planets is equal to the ratio of the cubes of their semi-major axes:

$$\frac{a^3}{a'^3} = \frac{T^2}{T'^2} = \frac{\omega'^2}{\omega^2}, \tag{2.1}$$

where a, T, ω denote the semi-major axis, the period and the angular velocity of a planet respectively. We can easily deduce that if a planet is at twice the distance of another, its angular velocity will be $\sqrt{8} = 2.8$ times less than that of the other.

The situation is similar in spiral galaxies. As a consequence of this, any extended object, such as an interstellar cloud, which rotates about the galactic centre tends to be deformed and stretched, as the parts closer to the centre rotate faster. Therefore, if the spiral arms were composed of the same matter they would have become like a string wrapped many times around the galactic centre. This is called the "coffee cup phenomenon", because it resembles the spiral patterns created on the surface of a cup of coffee when it is stirred. The coffee rotates faster in the centre than near the cup's rim, and the spiral becomes thinner and thinner until it disappears due to mixing. Consequently we would expect the spiral arms in

Fig. 2.6. The barred galaxy NGC 1365 (European Southern Observatory)

galaxies to disappear shortly after formation, and yet most galaxies have well defined spiral patterns that clearly have not been wrapped too many times around the centre. In the density wave theory, however, although the stars rotate faster near the centre the spiral pattern does not, and so its appearance does not change.

The creation of spiral arms is a very interesting problem. Various scenarios have been proposed. One obvious way is from the perturbation caused by the approach of another galaxy (*Toomre*). This approach creates two tidal waves, one in the direction of the other galaxy and one in the opposite direction. After the perturber has gone, the two tidal waves evolve into density waves and form the spiral arms. *Toomre* has made some impressive movies which show the creation of spiral arms by tides.

These encounters, however, are rather uncommon and so today spiral structure is believed to arise from instabilities in disc galaxies (*Lin, Shu,*

Kalnajs, Toomre etc.). It seems that matter concentrates in the centre of the disc in the form of a bar which triggers the creation of spiral arms.

The bar phenomenon is very general in galaxies (Fig. 2.6). It appears not only in barred spiral galaxies, where the bar is obvious, but also in ordinary galaxies, where we can sometimes find a small bar in the central regions. Also elliptical galaxies may be prolate and so be like bars. Bar formation can be easily explained theoretically, since the matter that collapses to form a galaxy is unlikely to do so perfectly symmetrically. The simplest and most common asymmetry is a bar-like asymmetry.

In the case of the barred spiral galaxies in particular, a simple explanation has recently been given for the formation of spiral arms. It appears that bars extend out to corotation, i.e. the radius where the rotational velocity of the stars matches the rotation of the bar. At smaller radii stars rotate faster than the bar, while at larger radii the stars rotate more slowly. In the region near to and outside corotation the stellar orbits are very irregular and lots of stars can escape to infinity. On the other hand, interstellar clouds collide with each other and form trailing spiral arms. This theory has recently been confirmed by numerical experiments concerning stellar orbits and orbits of individual clouds in barred galaxies.

As mentioned earlier, the spiral arms are areas of major concentration of interstellar matter where collisions between interstellar clouds are frequent. The collisions between interstellar clouds create shock waves along the spiral arms, which increase the local interstellar density sufficiently to cause collapse and creation of new stars. This is why the spiral arms are the regions where new stars form. These young stars ionise the hydrogen around them to create HII regions, which are so characteristic of the spiral arms. The young stars formed in the arms, later move out of them and spread around the whole galaxy. *Strömgren, Grosbøl* and others have calculated backwards in time the orbits of stars in our own Galaxy. Taking into account the ages of the stars and their current velocities they found that all the orbits converge to where the spiral arms were when the stars were formed.

2.6. Elliptical Galaxies

It is estimated that at least 30% of the galaxies are elliptical. These galaxies owe their name to their elliptical appearance when seen through a telescope (Fig. 2.7). Some of them are quite eccentric but others nearly round. They are roughly divided into two categories, giant ellipticals and dwarfs. There are, however, some intermediate types also, like the satellites of the Andromeda galaxy.

Ellipticals contain very little gas and even less dust. The existing gas is detected from radio observations in the 21 cm-line of neutral hydrogen.

Fig. 2.7. The elliptical galaxy NGC 205, a companion of Andromeda (Palomar Observatory photograph)

No formation of young stars is observed, and we do not see any blue Main Sequence stars. Consequently, elliptical galaxies are less bright than spiral galaxies. It appears that star formation occurred in a short period during the formation of these galaxies using up most of the available gas. Thus there was not enough gas left to form a disc with population I objects. In this respect elliptical galaxies are similar to globular clusters.

However, some gas remained, especially in the central regions. We often observe jets of matter coming out from the central regions of elliptical galaxies (Fig. 2.8), in contrast to their otherwise regular appearance. These jets emit strongly in radio wavelengths. They appear to be made up from gas ejected from the nucleus.

Detailed photometry shows that most elliptical galaxies have similar luminosity curves, which implies that they were formed by the same physical process. Both their colour and luminosity change very slowly with time.

To calculate the actual sizes of galaxies we must know their distances. Recent evidence shows a great spread in the diameters of elliptical galaxies, ranging from 1 to 150 kpc. The largest ellipticals are in the centres of

Fig. 2.8. The peculiar galaxy NGC 4486 (M 87). The radiation from the jet is polarised without any spectral lines and is synchrotron radiation (European Southern Observatory)

clusters of galaxies. They seem to have grown by "eating up" smaller galaxies that approached them and merged with them (Sect. 2.9).

2.7 The Local Group

Apart from the Magellanic Clouds and the Andromeda galaxy, there is a number of galaxies which lie in the vicinity of our own, within a distance of 1000 kpc, and compose the *Local Group* (Fig. 2.9) with approximately 30 known members. Seventeen are well studied and include 3 spirals (the largest ones), 10 ellipticals (mainly dwarfs) and 4 irregulars.

The diameter of the group is about 2000 kpc, with our Galaxy and Andromeda at two diametrically opposite points. One of the most distant members of the group is the Maffei II galaxy at a distance of 1000 kpc. The total mass of the group is about $5 \times 10^{12} \, M_\odot$.

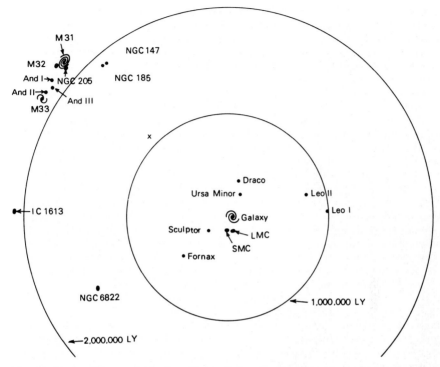

Fig. 2.9. The local group of galaxies with our Galaxy at the centre. The centre of mass of the local group is marked with an x

The Local Group moves as a whole towards a point in the direction of the Virgo cluster of galaxies, with a velocity of about 200 km/sec. (This velocity is to be subtracted from the general expansion velocity according to the Hubble law (Sect. 3.7), which is equal to 1100 km/sec.) The velocities of the individual members of the Local Group are not particularly high, so no member is believed to be able to escape the group, which is thus considered to be gravitationally bound.

2.8 Multiple Galaxies and Clusters of Galaxies

We often observe galaxies in pairs, like binary stars, or in small groups of three or more members. We also often see galaxies escorted by small irregular or elliptical companions, just as in the case of our Galaxy and Andromeda. The Sky Atlas of the Palomar observatory includes an estimated 100 000 multiple systems of galaxies. *Zwicky* estimates that 50% of all galaxies are in groups. An example of such a group is the Hercules

Fig. 2.10. The Hercules cluster of galaxies (Cerro Tololo Observatory)

cluster, shown in Fig. 2.10. Another example is the Virgo cluster which is about 20 Mpc from us, and has a mass of the order of 10^{15} $M_\odot$ with about 2500 known members. The Coma cluster has more than 1000 members and is at a distance of 40 Mpc.

Clusters of galaxies are held together by gravity, just like the star clusters. In both cases the principle is the same but the scale is different. There are also clusters of clusters of galaxies, called *superclusters,* which we shall talk about later (Sect. 3.2).

Observations have shown that when two galaxies approach each other they interact and as a result of the tidal effects, bridges of matter may form between them. These bridges are like huge arcs of matter connecting two or more galaxies. Such bridges have also been observed in the 21 cm-line. Sometimes, only one galaxy is visible with an *intergalactic tail,* the possible remnant of an old encounter with some passing galaxy. In multiple systems of galaxies we often observe a diffuse background of an *intergalactic medium,* which surrounds the system.

Some systems of galaxies have a double bridge connecting them, for example the galaxies NGC 5432–5435 (Fig. 2.11). In other cases (for example NGC 4038–4039) we have two filaments of matter, consisting of both stars and interstellar matter, starting from the point of contact of the two galaxies (Fig. 2.12). There are many similar cases of galaxies which are in contact and have suffered mutual deformation.

Fig. 2.11. Two galaxies connected by a double bridge (NGC 5432–5435) (Lick Observatory)

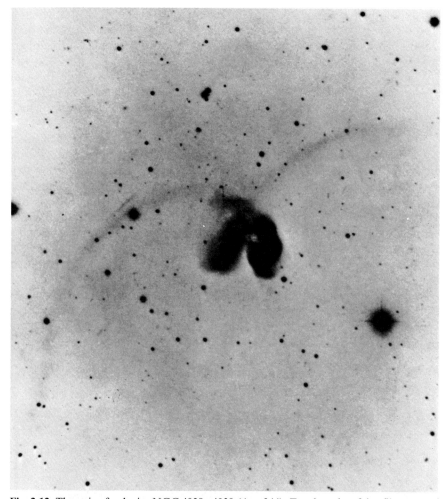

Fig. 2.12. The pair of galaxies NGC 4038–4039 (Arp 244). Two long but faint filaments of matter emerge from their point of contact (courtesy of Dr. Arp)

The spectra of the intergalactic bridges are similar to those of normal galaxies without emission lines. Theoretical work has shown that these bridges are the result of gravitational effects between the two galaxies, unlike the jets we mentioned in Sect. 2.6 which are probably due to phenomena occuring in the nuclei of the galaxies.

There is a lot of "intergalactic gas" between the galaxies. We can observe it directly from observations in the optical or in the radio band, but we can also infer its existence from indirect observations. We find, for example, that the total mass of a cluster of galaxies is greater than the sum of the masses of its members. We sometimes also observe, at radio wavelengths, wakes created behind moving galaxies (Fig. 2.13).

Fig. 2.13. Intergalactic gas heated up by a galaxy moving through it leaves a trace that radio astronomers can observe

The intergalactic matter possibly radiates because it is heated up by shock waves due to the fast motion of the galaxies. It is more likely, however, that the intergalactic matter has been heated up during the first stages of the cluster formation. The gas is very hot, with temperatures of hundreds of millions of degrees Kelvin, ten times higher than the temperature at the centre of the Sun. At this temperature the gas emits x-rays. The x-ray emission and the radio emission we receive from the intergalactic medium of a cluster is more than the radiation we receive from all the member galaxies together. Sometimes, however, a large part of the radiation we receive from a cluster comes from individual member galaxies like M 87 (of the Virgo cluster) and NGC 1275 (of the Perseus cluster).

2.9 Collisions of Galaxies

One of the most impressive phenomena in the Universe is a collision between galaxies. Such collisions occur mainly in the rich, compact clusters of galaxies where the density of galaxies is high. *Spitzer* and *Baade* estimated that in the Coma cluster every galaxy collides with 20 other galaxies every 3×10^9 years, assuming that the galaxies move mainly radially in the cluster.

The frequency of galaxy collisions may be even higher if one considers that every galaxy may be surrounded by a large dark invisible halo. These haloes could be sufficiently large to span the distance between the galaxies. The implication is that galaxies often move inside the outer parts of other galaxies. When a galaxy moves inside the halo of another galaxy it is slowed down by what is called "dynamical friction". As a result the two galaxies approach each other even closer and eventually the smaller of the two is absorbed by the larger. This phenomenon is referred to as "galactic cannibalism".

Another type of galaxy interaction is the approach of two galaxies which move fast with respect to each other. The two galaxies pass through each other without their stars colliding. Indeed, the volume of space inside each galaxy is so large that in the eventuality of such a head on collision the probability of one star colliding with another is very small. Most of the stellar orbits remain essentially unaltered and only a few stars acquire escape velocities. However, the interstellar matter of the two galaxies collides strongly, and forms a large cloud of galactic dimensions, while the two galaxies, which are now clean of gas, move away in opposite directions. This phenomenon explains some cases of intense radio emission due to the collision of the interstellar matter of two galaxies.

Finally, two galaxies may approach each other with intermediate relative velocity. In this case the two galaxies affect each other significantly and change their shapes. Tidal effects may induce spiral arms which will persist after the two galaxies have separated. Also, bridges of matter are created between the galaxies and a tail of matter may get pulled out on the opposite side (Figs. 2.11, 2.12). The phenomenon of galactic approaches and collisions is particularly impressive when observed in simulated movies from numerical experiments (*Toomre, Miller* and others). One can see the formation of bridges and tails like those we observe in irregular galaxies and galaxy pairs.

2.10 Distances of Galaxies and Clusters of Galaxies

The following methods are used today to estimate the distances of galaxies and clusters of galaxies:

1) In some galaxies we observe cepheid variable stars. From the period-luminosity relation which holds for such stars we may find the absolute magnitude M. If m is the apparent magnitude of such a star its distance r in pc is given by:

$$m - M = 5 \log r - 5. \tag{2.2}$$

This method has been applied for distances up to 4 Mpc.

2) For more distant galaxies, where we cannot observe cepheid stars, we use as standard candles the novae and supernovae. We consider that the absolute magnitude of a supernova is known and from its observed magnitude we can again estimate its distance using the formula (2.2). Such a method can be applied for distances up to 100 Mpc. Alternative standard candles are the globular clusters of the galaxies.

3) For the most distant galaxies and clusters of galaxies we use the Hubble law which relates the redshift of the spectrum of a galaxy to its distance from us. The shifting of a spectral line of wavelength λ, by $\Delta\lambda$ towards the red part of the spectrum of a galaxy, is related to the galaxy's recession velocity v by the formula:

$$v = c \frac{\Delta\lambda}{\lambda}, \tag{2.3}$$

where c is the speed of light. According to an approximate expression of Hubble's law, if r is the distance of a galaxy from us, its recession velocity is

$$v = Hr, \tag{2.4}$$

where H is a constant, which is called Hubble's constant.

By combining the above two equations we may estimate the distance of a galaxy. More details about this subject are given in Sects. 3.7, 6.4 and 6.5.

2.11 Masses of Galaxies and Clusters of Galaxies

The masses of galaxies may be estimated by various methods. For nearby galaxies a reliable method is based on the rotation velocity of the galaxy. If the rotation velocity of a spherical galaxy is v at a distance r from its rotation axis, by equating the centrifugal force

$$\frac{v^2}{r} \qquad (2.5)$$

to the gravitational force

$$\frac{G M(r)}{r^2}, \qquad (2.6)$$

where G is the gravitational constant and $M(r)$ is the mass of the galaxy within radius r, we obtain the following expression for the mass of the galaxy:

$$M(r) = \frac{r v^2}{G}. \qquad (2.7)$$

As r tends to the limiting radius of the galaxy, $M(r)$ tends to its total mass. The rotational velocity $v(r)$ of a galaxy can be calculated from the redshift of the 21 cm-line of neutral hydrogen (Fig. 2.14). By this method it has been found that the masses of spiral galaxies are of the order of $10^9 - 10^{12} \, M_\odot$.

Another way of estimating the mass of a galaxy is from the distribution of the velocities of the stars inside the galaxy, by using the Virial theorem (Sect. 2.12).

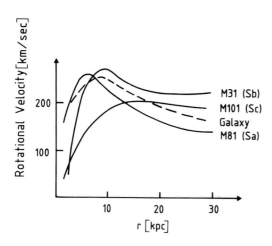

Fig. 2.14. The rotation velocity as a function of distance from the centre, for various galaxies

In the cases of pairs of galaxies we can apply the methods used to calculate the masses of double stars. We equate the centrifugal force to the gravitational force, using the velocities of the two galaxies with respect to their centre of mass. This technique yields the sum of the two masses.

If we know the mass to luminosity ratio for a cluster, we may estimate the total mass of the cluster from the luminosity distribution.

The Virial theorem (Sect. 2.12) can be applied to a cluster of galaxies by considering the radial velocities of the member galaxies. This technique yields masses of the order of $10^{13} - 10^{15}\ M_\odot$.

An alternative method for calculating the mass of a cluster is to add up all the individual masses of the member galaxies. The cluster mass calculated in this way turns out to be much lower than that estimated using the Virial theorem. This is why we talk about "missing mass" in clusters of galaxies.

There are two possible explanations for the missing mass problem. Either the Virial theorem does not apply to clusters of galaxies, or there is a lot of invisible mass in the clusters. The first case applies when a cluster is expanding fast. In such a case the cluster will quickly disperse, as the gravitational attraction of the individual members is insufficient to hold them together. We therefore, prefer the second alternative. The invisible mass could consist of faint dwarf stars or black holes, but in recent years it has been suggested that it may consist of "heavy neutrinos" or other exotic particles (Sect. 6.6).

Since we can calculate the masses of galaxies and clusters of galaxies we may find the average density of the Universe, because we can estimate the number of galaxies and clusters in a given volume. The calculation of the average density of matter in the Universe is a very important and difficult problem. From the existing data on the distribution of galaxies and clusters of galaxies in space, and their masses, we may find an *average density of visible matter in the Universe,* which turns out to be of the order of 2×10^{-30} g/cm^3.

The correct estimate of the average density of matter in the Universe has cosmological implications, because it determines the curvature of the Universe, which in turn is crucial in determining whether the Universe is closed (finite) or open (infinite) (Sects. 6.2, 6.6).

On average we observe 0.02 galaxies per cubic megaparsec. Table 2.1 summarises the most important characteristics of galaxies based on the most recent observational data.

Table 2.1. Characteristics of galaxies

	Spirals	Ellipticals	Irregulars
Mass (in solar units)	$10^9 - 10^{12}$	$10^6 - 10^{13}$	$10^8 - 10^{11}$
Diameter (kpc)	$6 - 50$	$1 - 150$	$1 - 10$
Luminosity (in solar units)	$10^8 - 10^{10}$	$10^6 - 10^{11}$	$10^7 - 10^9$

2.12 The Virial Theorem

The Virial theorem tells us that the sum of the potential energy and twice the kinetic energy of a self-gravitating system is zero:

$$E_{pot} + 2E_{kin} = 0. \tag{2.8}$$

This formula holds for systems in statistical equilibrium, i.e. systems which do not change appearance as time passes. If R is the radius of a cluster (assumed spherical) within which half of its mass lies, we have the following formula for its potential energy:

$$E_{pot} = -\frac{GM^2}{2R}, \tag{2.9}$$

where M is the total mass of the cluster and G is the gravitational constant. The kinetic energy of the system is given by

$$E_{kin} = \tfrac{1}{2} M \langle v^2 \rangle, \tag{2.10}$$

where $\langle v^2 \rangle$ is the mean square velocity of the cluster members about the cluster centre. From the above formulae we obtain the total mass of the cluster:

$$M = \frac{2 \langle v^2 \rangle R}{G}. \tag{2.11}$$

The velocities of the member galaxies can be found from spectroscopic observations, because of the Doppler effect (shifting of the spectral lines), by using the formula (2.3).

The above method yields a mass of $8 \times 10^{14} M_\odot$ for the Coma cluster. Given that there are roughly 800 member galaxies, the average mass of each member turns out to be $10^{12} M_\odot$, much higher than the observed average mass of the discs of galaxies.

2.13 Peculiar Galaxies

Although the "normal" inhabitants of the Universe are the galaxies, not all galaxies are "normal". The nuclei of some galaxies suffer extremely violent events, which affect the whole galaxy. The major pecularity of these galaxies is that they radiate thousands of times more energy in the radioband than the normal galaxies do.

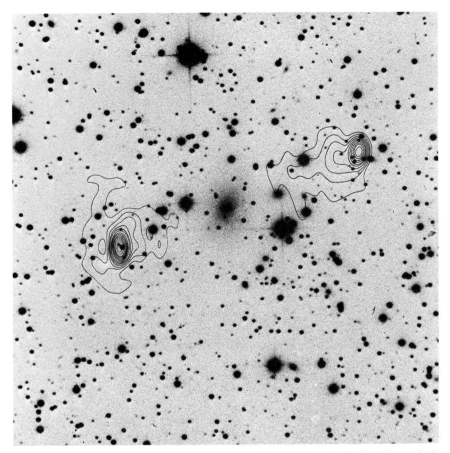

Fig. 2.15. The strong radio source Cygnus A, one of the brightest in the sky. The optically visible galaxy is at the centre but the radio emission comes from two regions which cannot be seen in the optical

With radio telescopes we can observe the Universe in wavelengths from 1 mm to 100 m and make detailed radio maps of the sky. Some of the sources in these maps are stars and nebulae, while others are galaxies. Most of the radio sources that turn out to be galaxies are rather weak. The Andromeda galaxy, for example, emits 5×10^{38} erg/sec in the radioband, while our own Galaxy is thought to emit approximately 10^{37} erg/sec in the same band. Other radio galaxies include the Magellanic Clouds and M 51 (Fig. 2.4). We also receive radio waves from many clusters of galaxies.

Some radio galaxies, however, which are very far from us, appear to be particularly powerful. Some of them have a peculiar appearance, while others look like normal galaxies. These powerful radio sources are called *peculiar galaxies*. They seem to have in their nuclei powerful sources of energy which emit radiation either continuously or in explosions.

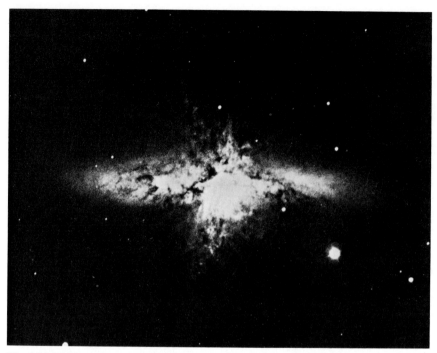

Fig. 2.16. The irregular galaxy M 82 with many filaments coming from the nucleus. It is also a strong source of radiation in the infrared, x-rays and radio waves (Palomar Observatory photograph)

The release of energy appears in the form of a cloud of high-energy protons and electrons, moving along magnetic field lines. Although we do not know what causes the explosion, we can describe in detail the synchrotron radiation we receive from the electrons. These electrons are relativistic (with velocities approaching the speed of light) and are emitted from an extended region. Many radio galaxies have two emission centres, separated by a distance greater than 100 kpc and located symmetrically about the optically visible galaxy. An example of such a radio source is Cygnus A (Fig. 2.15) which is 220 Mpc away from us. In general such strong radio sources emit a power in excess of 10^{45} erg/sec.

Initially it was thought that the radio source of Cygnus A, along with other radio sources, was the result of a galactic collision. During such a collision, the kinetic energy of the interstellar matter is suddenly converted into thermal energy and thus radiates strongly at radio wavelengths. However, the energy radiated in this way is inadequate to explain the most powerful radio sources we observe. Also, if we accept the present distance scale to the galaxies, the frequency of galactic collisions is much less than the frequency of observed radio sources in the Universe. Finally, this theory predicts radio emission from one source and not from two separate centres.

The most likely explanation is that the intense radiation we receive from these galaxies is due to violent events which occur in their nuclei, such as a powerful explosion, or the presence of a massive black hole which is continuously swallowing stars.

An example of a powerful radio source is the peculiar galaxy M 82, which is also an x-ray source and a strong infrared source (Fig. 2.16). Its filamentary appearance is believed to be the result of an explosion in its nucleus. The filaments in the outer parts of the galaxy move with a velocity of a few hundred km/sec, but the velocities with which the electrons move in the central regions are believed to approach the speed of light. Another intense radio source is the radio galaxy Virgo A which coincides with the galaxy M 87. This galaxy has an impressive jet of matter ejected from its nucleus (Fig. 2.8). Its light is blue, continuous and polarised without emission lines. Again it is synchrotron radiation.

2.14 Quasars

The quasars are a separate category of radio galaxies, which are particularly active. They are the most distant objects in the Universe that we can see, at distances of billions of light years from us. They owe their name to their stellar appearance when seen through a telescope (Fig. 2.17). The first quasar was found in 1960 with the Palomar telescope, and is called 3C48 (the 48th radio source listed in the third Cambridge catalogue). It was observed as a faint 16th magnitude star, but its emission lines did not look like the emission lines of any known star. So, this star was a mystery. In 1963 a second star with similar spectrum was discovered at the position of the radio source 3C273. At that time *Schmidt* thought of measuring the *ratios* of the wavelengths of the most intense lines of the spectrum and found that these ratios were identical to the ratios of the hydrogen lines. He immediately realised that the emission lines observed in the spectra of the quasars are the well known hydrogen lines strongly redshifted. The relative redshift was found to be 16%, i.e. the velocity with which the source recedes from us is 15% of the velocity of light. Later on it was established that the spectrum of 3C48 was an ordinary spectrum too, redshifted by 37% which corresponds to a recession velocity of 30% that of light.

These discoveries prompted the astronomers to look for more quasars (Fig. 2.17). Indeed, more than 800 quasars are now known with redshifts $z = \Delta\lambda/\lambda$ up to $z = 4$ which corresponds to recession velocities of 92% the speed of light. (When the velocities are large, we must use the relativistic *Doppler* formula which relates the velocity of the source to the redshift (Sect. 6.4).)

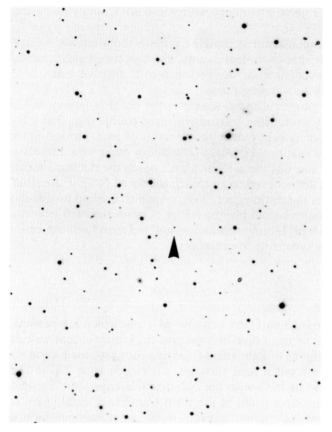

Fig. 2.17. The quasar 3C9, appears as a faint star among other stars (Palomar Observatory photograph)

It is generally accepted that quasars are at cosmological distances and that their great velocities are due to the expansion of the Universe (Sect. 6.4). They are, therefore, the most distant objects in the Universe we can observe. The closest quasar (3C273) has a redshift of 0.15. A confirmation of the idea that quasars are very distant objects in the Universe is that in certain cases a galaxy has been detected surrounding the quasar (implying that the quasar is the nucleus of the galaxy), and the distance of that galaxy has been estimated. Further, it has been observed that some quasars belong to groups of galaxies which have the same redshift as the quasar. Even the quasar 3C273 which so far had been thought of as a rather peculiar one, because it appeared isolated, was recently shown to be in a small group of galaxies with similar redshifts.

A problem stressed by several people is that quasars have enormous luminosities, 100 or more times greater than the luminosity of our Galaxy.

That is why some scientists have suggested that quasars have been ejected from relatively closeby galaxies. *Arp* in particular has found cases of quasars appearing near relatively closeby galaxies. However, if *Arp*'s suggestion is correct the problem is not really solved, because one has then to explain how the quasars are ejected with such high velocities. This also does not explain the fact that all observed quasars are receding from us and no quasar has been ejected towards our direction. For all the above reasons, most scientists accept that quasars are at cosmological distances.

A new confirmation of this view is the discovery of the "gravitational lenses". If a galaxy is located between the Earth and a distant quasar, it is possible for two or more quasar images to be seen. This phenomenon is due to the same effect that causes the path of the light from stars to bend as it passes close to the Sun, i.e. the presence of a strong gravitational field.

Two cases of gravitational lensing have been discovered in 1979 and 1980. The first one concerns the double quasar 0957-561. The two quasars are characterised by the letters A and B and have a separation of 6 arc secs. Quasars are not very numerous, so the discovery of two of them at such a small separation from each other is very unlikely. Soon after their discovery it was found that both quasars have the same redshift and entirely similar spectra. This convinced astronomers that they were looking at the double image of the same quasar created by a gravitational lens. Later a faint galaxy was found between the two images, which is thought to produce the bending of the light from the quasar. Until the beginning of 1984 a total of 5 gravitational lenses had been identified.

The above examples indicate that quasars are really very distant galaxies. Therefore, they are particularly interesting objects for two reasons: (1) because they reveal how galaxies were billions of years ago, when the light we receive now first set off from these quasars, and (2) because their distribution reveals the density of the Universe at that time (Sect. 6.8).

Many theories have been developed to explain what powers the quasars. Some people suggest that they are very faint rotating magnetic superstars with masses $10^8 - 10^{10} \, M_\odot$. It has also been suggested that there could be many supernovae simultaneously triggered by multiple collisions of stars in the galactic centre. The best theory seems to be one based on a large black hole at the centre of the quasar (*Lynden-Bell*). The energy of the quasar is produced by the continual supply of stars falling into it.

Besides the quasars, there are other galaxies with active nuclei, like the Seyfert galaxies (named after the astronomer who discovered them), the BL Lacertae objects, and various peculiar galaxies. Many Seyfert galaxies are also radio sources as well as strong sources in the ultraviolet, the infrared and the x-rays. Their infrared luminosity is sometimes 100 times the optical luminosity from our Galaxy. The BL Lacertae objects are like quasars without emission lines. This lack of emission lines is attributed to the absence of gas around them. All these objects are studied intensively nowadays.

3. Distribution of Matter in the Universe

3.1 Intergalactic Matter

The clusters of galaxies have a diameter greater than 1 Mpc. The member galaxies have velocities of the order of 1000 km/sec and require a billion years to cross them. In some cases the galaxies require even more time to cross the cluster once, a good fraction of the age of the Universe. Even so, the velocities of galaxies inside a cluster are much greater than the velocities of stars inside a galaxy. For example, the Sun moves with a velocity of 220 km/sec about the galactic centre, while the galaxies in the Coma cluster move with velocities of 2000 km/sec. This could mean that some groups and clusters of galaxies may be unstable, having insufficient mass to hold the member galaxies together. It is more likely, however, that there is a lot more matter in a cluster than we can observe. It must be in the form of some intergalactic medium, and there must be sufficient of it to prevent the member galaxies from escaping. We have two indications that this is true: (1) The velocity curves of galaxies show that galaxies are surrounded by massive halos. (2) We frequently observe matter between the galaxies.

As interstellar matter spreads between stars, so too intergalactic matter spreads between galaxies and between clusters of galaxies. It consists mainly of hydrogen, but it also contains helium and other heavier elements in smaller proportions. It can be detected by radio observations in the 21 cm-hydrogen line. It is also made manifest by the "trace" left behind a galaxy as it moves through a cluster (Fig. 2.13). The existence of intergalactic matter is further confirmed by direct observations of luminous matter, e.g. the bridges joining together some galaxies. The study of the distribution of the intergalactic medium shows that it is present everywhere in intergalactic space. There is probably very little intergalactic dust, making up 1% of the whole, as in the case of the interstellar dust. The intergalactic dust probably consists mainly of large snow flakes of solid hydrogen.

The rich clusters of galaxies seem to have at least as much diffuse intergalactic matter as visible matter in the form of galaxies. Some theories of galaxy formation require the existence of a relatively dense intergalactic gas ($\simeq 10^{-29}$ g/cm^3) of high temperature (10^8 K). This idea is supported by x-ray, ultraviolet and radio observations.

The intergalactic gas was probably heated during the formation of clusters of galaxies and subsequently became so diffuse that it could not cool down. Its temperature is 10^8 K and so it is expected to emit x-rays. Indeed, the Uhuru satellite has detected x-rays coming from the rich clusters of galaxies in Virgo, Coma, Perseus and Centaurus. Subsequent observations with Ariel 5 showed that the emission of x-rays is a property of *all* clusters and that the amount of radiation depends upon the number of galaxies and their compactness in the cluster. All these observations have to be done by satellites since the Earth's atmosphere entirely absorbs x-rays so that they never reach the surface of the Earth.

The x-ray spectrum of the intergalactic gas has at least one emission line which is produced by nuclei of ionised iron. Observations with Ariel 5 show a clear increase in the received radiation at wavelength 1.77 Å ($= 1.77 \times 10^{-8}$ cm). This line is due to atoms of iron which have lost all but one or two electrons (the neutral iron atom has 26 electrons). The existence of this line has been the major support for the suggestion that the x-ray radiation is of thermal origin, due to gas at temperatures of 10^8 K. Such temperatures, 5–10 times higher than the temperature in the centre of the Sun, are necessary to ionise the iron to the degree we observe.

Three more mechanisms have been proposed for the heating of the intergalactic gas in clusters. The first argues that supernovae continuously enrich the intergalactic medium with hot gas emitted away from a galaxy in the form of a galactic wind. The second mechanism proposes that close encounters of two member galaxies cause the heating of the gas. The third mechanism has to do with the stripping of gas from the galaxies as they move through the cluster. This last mechanism preassumes that a large quantity of intergalactic matter already exists.

The intergalactic gas is most likely the remnant of the "primordial gas" that collapsed gravitationally and formed the clusters of galaxies. Indeed, during the galaxy formation era, not all the gas is expected to have been concentrated inside galaxies; a large part of it must have been dispersed throughout space, as is known to happen during the formation of stars from the interstellar medium.

The iron observed in the intergalactic medium is in the same proportion to hydrogen as that observed in the Sun. This is particularly surprising since, as is known, iron, as well as other heavy elements, are created in the stellar interiors and spread throughout space after the stars explode at the end of their evolution. Since we expect the intergalactic medium to be of primordial origin, such an observation has important consequences. A possible explanation is that iron was formed by a first generation of stars (population III, Sect. 8.3) before the formation of galaxies, and was subsequently spread in space by supernova explosions.

An alternative explanation is that this gas is not of primordial origin after all, but comes from the galaxies, just as part of the interstellar medium comes from stars. For a start, there is the galactic analogue of the solar wind, called the galactic wind. Further, there is evidence that some galactic nuclei eject gas which not only spreads inside the galaxies themselves, but also into intergalactic space. Such phenomena are more intensive in Seyfert galaxies, which include in their spectra emission lines due to hot gas. These galaxies are sources of high-energy cosmic rays. Some of them have jets of matter which move with speeds of a few thousand kilometres per second. For example, the galaxy NGC 4151 emits from its nucleus three jets of gas which move with velocities of 280, 500 and 840 km/sec, with the result that the galaxy loses $10-1000\ M_\odot$ per year. The radiation we receive from such galaxies is non-thermal synchrotron radiation. This implies that there are relativistic electrons moving inside magnetic fields.

3.2 Superclusters

Shapley and *Ames* published a catalogue of the positions of bright galaxies in 1932. It became clear from this catalogue that galaxies segregate into compact clusters, some of which are larger than the Virgo cluster and several appear spherically symmetric. *Abell* in 1958 chose 1682 clusters from the catalogue, which formed a homogeneous sample. He noticed that apart from the clusters of galaxies there are also "superclusters" due to second order clustering. That is, he noticed that the rich clusters have a tendency to segregate into groups of clusters of galaxies. The sizes of these superclusters are of the order of 50 Mpc. According to *Abell* and *de Vaucouleurs*, there are superclusters which include 10 rich clusters and have diameters 40 Mpc and masses $10^{15}-10^{17}\ M_\odot$. The largest superclusters could have as many as 100 000 member galaxies.

De Vaucouleurs found evidence that there is a "cluster of clusters of galaxies" around our Galaxy, called the "local supercluster" (Fig. 3.1). It is a flat ellipsoidal system of 15 Mpc cross-section and 1 Mpc thickness, which also includes the local group of galaxies. Our Galaxy is somewhere near the edge of this "metagalactic system". *De Vaucouleurs* found that this system rotates and expands, by studying the radial velocities and dynamics of 50 000 member galaxies. From its rotation he estimated that its total mass is $10^{15}\ M_\odot$. The whole local supercluster (including our Galaxy) moves with respect to other distant superclusters with a velocity of the order of 500 km/sec.

There are also other ways of finding the form and degree of clustering of galaxies. *Peebles* and others have studied the distribution of galaxies using the information from various catalogues of galaxies and clusters of galaxies, and calculated the autocorrelation function for this distribution. The qualitative and quantitative study of these data led *Peebles* and others to conclude that galaxies form clusters of higher and higher dimensions. Thus a hierarchy of clusters is formed.

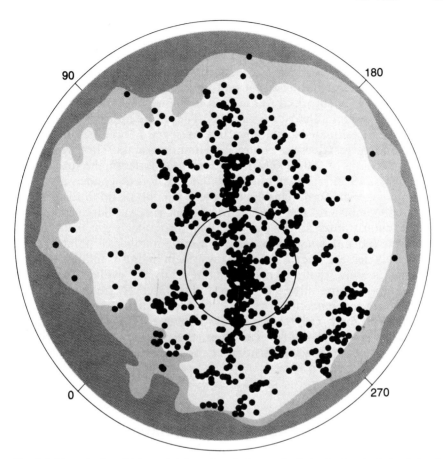

Fig. 3.1. The galaxies which are brighter than 13th magnitude in the North Galactic hemisphere. The North Galactic Pole is at the centre. Inside the small circle is the Virgo cluster which is part of the local supercluster and is at a distance of 20 Mpc. The boundary represents the galactic equator, where the absorption due to the dust in our Galaxy is greatest

It is estimated today that 90% of the galaxies belong to clusters and superclusters. The superclusters are flat, and for this reason they are sometimes called "pancakes". Between them there are large spaces almost empty of any galaxies, the so called "voids".

3.3 Voids

At first glance the observations of clusters and superclusters indicate that their distribution is random. In recent years, however, people have collected a large amount of data on the distances of galaxies, and thus it became possible to study their distribution in three-dimensional space. (The distances of galaxies and clusters of galaxies can be found by the methods we described in Sect. 2.10.) When one considers the positions of galaxies in space by taking into account their distances from us, it becomes clear that their distribution is not uniform.

Scientists were surprised to find that there are huge areas in the Universe which contain almost no galaxies. They were appropriately called "voids". It is estimated that only 10% of space is occupied by superclusters and the rest does not contain any luminous mass. These voids may reach dimensions of 100–200 Mpc. The recent theories on the structure of superclusters (Sect. 8.3.1) can explain the creation of condensations of matter in spherical form (clusters of galaxies mainly), and in filamentary and flat superclusters. The various simulations run with large computers are particularly interesting. The computer program is given the initial conditions (positions and velocities) of a large number of points, representing galaxies, and the evolution of the system in time is followed. These experiments show that the matter in the Universe evolves to form a honeycomb-like structure. The cells of the honeycomb are the voids and the walls are the superclusters. It is estimated that there are about one million honeycomb cells in the visible Universe. This structure of the Universe is slowly changing. Therefore, the present appearance of the Universe reflects the whole history of the formation and evolution of superclusters.

3.4 Isotropy and Homogeneity of the Universe

The Universe appears to be isotropic and homogeneous on large scales. When we say that the Universe is isotropic, we mean that an observer will see the same characteristics in the Universe whichever direction he observes. When we say that it is homogeneous, we mean that the Universe will appear the same to any observer, independently of his position. In other words, all observers, wherever they are, will find the same density and generally the same properties of the Universe. They will, therefore, form the same picture of it. This principle, according to which the Universe is isotropic and homogeneous, is called the "cosmological principle".

Let us see, however, to what degree these two properties characterize the actual Universe. At first sight the distribution of galaxies does not appear at all regular (Fig. 3.2). A strip close to the galactic plane does not include any galaxies, while at latitudes of $10°-40°$ the distribution is irregular, as the number of galaxies we see increases with galactic latitude. We know today that the absence of galaxies close to the galactic plane is due to interstellar absorption in our own Galaxy. The light of galaxies at

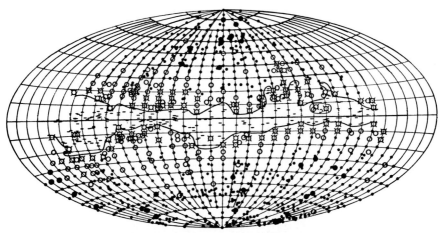

Fig. 3.2. The distribution of galaxies in galactic co-ordinates down to 20th magnitude. Close to the galactic plane there is a zone of avoidance, where no galaxies are observed. The circles represent more than one galaxy (∘∘∘: small numbers of galaxies; ●●●: large numbers of galaxies)

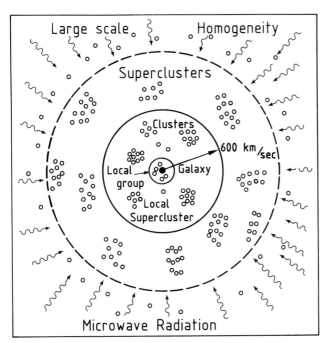

Fig. 3.3. The distribution of galaxies is inhomogeneous on small scales, but becomes increasingly homogeneous as the scale increases. The microwave background radiation has the greatest degree of homogeneity

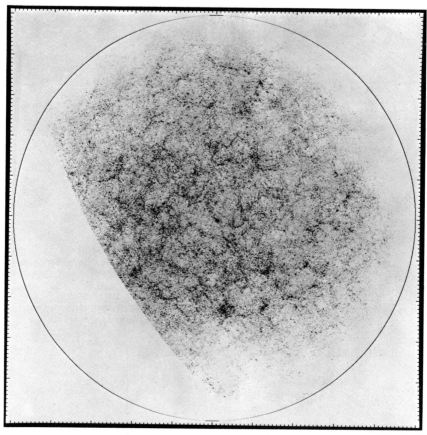

Fig. 3.4. Map of the galaxies in the North galactic hemisphere including all galaxies brighter than 19th magnitude (there are about one million of them). The North Galactic Pole is at the centre while the circle is the galactic equator. The darker regions represent more galaxies. The white region on the left cannot be seen from observatories in the Northern hemisphere of the Earth. The absence of galaxies close to the equator is due to absorption by our Galaxy. The segregation of galaxies into clusters and superclusters is apparent as well as the homogeneity of the distribution on large scales

small latitudes is absorbed by the large amount of interstellar matter near the galactic plane.

If we make the correction for the interstellar absorption, the apparent distribution of galaxies becomes *isotropic*. That is, in any direction we observe, the number of galaxies per unit solid angle is constant.

A proof for the homogeneity of the distribution of galaxies is based on the observation that the number of galaxies up to magnitude $m + 1$ is about four times the number of galaxies up to magnitude m, as is expected if the density of galaxies in space is constant (Sect. 3.5).

The statistical analysis of the distribution of galaxies shows a strong tendency for galaxies to segregate into clusters of 50 or more members. The clusters of galaxies segregate into superclusters. As we proceed to larger and larger scales, the homogeneity of the Universe becomes clearer (Fig. 3.3).

The first observations of the distribution of galaxies and clusters of galaxies in the Universe by *Shapley* and *Ames* (1932) did not find homogeneity because their observations were restricted to a relatively small region.

Abell (1958), in his study of the distribution of clusters and superclusters in space did not find any systematic difference in the distribution of clusters in the northern and southern hemispheres, nor did he find any systematic change in their distribution as a function of distance. Thus he concluded that there is large scale isotropy and homogeneity. The most recent studies by *Abell* and others (1982) are consistent with the view that the Universe is homogeneous on scales of the order of 1000 Mpc (Fig. 3.4).

The strongest evidence for the isotropy and homogeneity of the Universe comes from the observations of the "microwave background radiation". This radiation is the remnant of the first stages of the evolution of the Universe, and we receive it from the extremities of the Universe. It comes with exactly the same intensity from all directions (Sect. 3.6). The possible anisotropy is smaller than 10^{-4} (one part in ten thousand).

3.5 Counts of Galaxies

The total number of galaxies brighter than apparent magnitude m is N_m:

$$N_m = \tfrac{4}{3} \varrho \pi r_m^3, \tag{3.1}$$

where r_m is the distance from the Sun of a galaxy of apparent magnitude m, and ϱ is the number of galaxies per unit volume, which is assumed constant. Similarly, the number of galaxies brighter than magnitude $m + 1$ is

$$N_{m+1} = \tfrac{4}{3} \varrho \pi r_{m+1}^3. \tag{3.2}$$

Therefore we have the relation

$$\frac{N_{m+1}}{N_m} = \left(\frac{r_{m+1}}{r_m}\right)^3. \tag{3.3}$$

The apparent magnitude of a galaxy of certain absolute luminosity varies inversely proportionally to the square of its distance. So, a galaxy at a distance r_m is brighter than another one, with the same absolute luminosity, at a distance r_{m+1}, by a factor of $(r_{m+1}/r_m)^2$. It is known, however, that the luminosity ratio of two objects of magnitudes m and $m+1$ is $10^{2/5}$, that is $(r_{m+1}/r_m)^2 = 10^{2/5}$. (Indeed, by definition, objects of first magnitude are 100 times brighter than objects of 6th magnitude; if the 1st magnitude objects are α times brighter than the 2nd magnitude objects, which are α times brighter than the 3rd magnitude objects, and so on, it turns out that $\alpha^5 = 100$, i.e. $\alpha = 10^{2/5}$.)

Therefore Eq. (3.3) takes the form

$$\frac{N_{m+1}}{N_m} = 10^{3/5} = 3.98 \simeq 4. \tag{3.4}$$

That is, the total number of galaxies up to magnitude $m+1$ is four times greater than the number of galaxies up to magnitude m. This is, in fact, what we observe, and justifies our assumption that the number density of galaxies is constant throughout space.

3.6 Microwave Background Radiation

The microwave background radiation was discovered in 1965 by *Penzias* and *Wilson* and was one of the major cosmological discoveries of all times. This is the remnant of the radiation of the early Universe, and uniformly fills the whole of space. It is one of the most important indications that the Universe comes from an initial state of very high density and temperature, called the Big Bang.

The discovery by *Penzias* and *Wilson* was essentially fortuitous. That is, these observers did not set out to observe this radiation, although its existence had already been predicted by *Gamow* and his collaborators in 1946–48, as a part of their theory for the formation of the chemical elements. *Penzias* and *Wilson* were simply studying the distribution of microwaves which come from space, at wavelengths of a few centimetres. They found that there was a diffuse radiation which was coming uniformly from all directions and corresponded to a black-body spectrum of approximately 3 K. This radiation was unexpected because it could not be explained by the known terrestrial or extraterrestrial sources such as the Sun or the stars. Then they received the information that *Dicke* and *Peebles* at Princeton had examined the possibility of observing the radiation predicted by *Gamow* and actually *Dicke* and his collaborators were building a special instrument to observe it. So, when in 1965 *Penzias* and *Wilson* published their observations, the astronomers at Princeton published at the same time the interpretation of these observations.

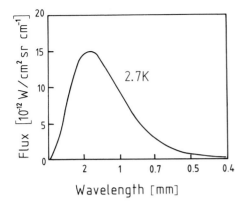

Flux [10^{12} W/cm^2 sr cm^{-1}]

2.7K

Wavelength [mm]

Fig. 3.5. The spectrum of the micro-wave background radiation according to recent observations. The continuous line represents the radiation flux for a black body at a temperature of 2.7 Kelvin, in units of 10^{-12} Watts per cm^2 per unit solid angle (sterad) per unit of inverse wavelength (cm^{-1})

Immediately afterwards a systematic study of this microwave radiation commenced. The first characteristic of this radiation was its almost absolute isotropy. That is, the differences in the intensity of this radiation from one point in the sky to the next have been found to be less than 10^{-4}, i.e. less than 0.01%. The implication of this is that this radiation is not due to stars or galaxies, but to the early concentration of matter in the Universe when its temperature was about 3000 K (Sect. 8.1). Proof for the origin of this radiation comes from the fact that its spectrum is very nearly that of a black body (Fig. 3.5). The first observations were made at relatively long wavelengths (left side of the curve of Fig. 3.5) and they did not prove that it had a black-body spectrum. Since then, however, more observations have been made and it has now become generally accepted that the spectrum of the microwave background radiation corresponds, to high accuracy, to a black-body spectrum of temperature 2.7 K.

Some more recent observations (1980–81) have shown that there is a small anisotropy in the microwave radiation. This anisotropy has an amplitude of 10^{-3} and is bipolar, that is the intensity has a maximum in one direction and a minimum in the opposite direction. This can be easily explained if we accept that the Earth and the Galaxy move, with respect to the distant galaxies of the Universe, with a speed of 600 km/sec, towards a certain point in the sky (with galactic co-ordinates $l = 260°$ and $b = 35°$; Fig. 3.3). This velocity and co-ordinates agree qualitatively with independent observations concerning the motion of our Galaxy with respect to remote galaxies (Sect. 3.2).

In 1981 a small quadrupole anisotropy was reported (i.e. the radiation seemed to have maxima in two opposite directions). These observations, however, have not been confirmed by more recent studies (1982). Some anisotropy with an amplitude of the order of 10^{-4} is expected, due to the existence of clusters of galaxies in the Universe. Therefore, observations of anisotropies in this radiation will provide significant information on the distribution of matter on all observable scales.

3.7 Expansion of the Universe

Shlipher in 1914, working at the Lowell observatory, was the first to notice that the spectra of 40 or more galaxies he had observed, apart from a few exceptions, were redshifted. This can be explained if we accept that these galaxies move away from us with large velocities. The few exceptions (like the Andromeda galaxy) are due to the fact that the Sun moves around the centre of the Galaxy with a velocity of 220 km/sec, thus the nearest galaxies in the direction of its motion seem to approach us. Later on, *Hubble* and *Humason* at the Mount Wilson observatory, found the same effect in many other galaxies which were at greater distances. The same phenomenon has been observed in clusters of galaxies, radiogalaxies and quasars. These observations have been carried out using the large telescopes at Palomar, Lick and Kitt Peak observatories, and the observatories in the southern hemisphere. The recession velocities of galaxies were measured, and the largest were found to be those of the quasars, reaching hundreds of thousands km/sec. For example, the quasar 3C9 (Fig. 2.17) recedes with a velocity 0.8 that of light, that is 240000 km/sec. Also the quasar OQ172 has a redshift corresponding to a recession velocity of 91% the speed of light, i.e. 273000 km/sec.

The further away the galaxy we observe is, the less important the contribution of the galaxy's proper motion to its recession velocity becomes. In general, the recession velocities of galaxies increase with their distance from us (Fig. 3.6).

In 1929 *Hubble* deduced the law which relates the velocity of a galaxy to its distance:

$$v = Hr,$$

where v is the radial velocity of the galaxy in km/sec, r the distance of the galaxy in Mpc and H is a constant called Hubble's constant. That is, the recession velocity of a galaxy is proportional to its distance. The determination of the exact value of the constant H is very important because it is related to the question of which theoretical model of the Universe is more appropriate. For this reason a lot of effort has been put into an accurate determination of H to find the exact velocity-distance relation for distant galaxies.

The value of Hubble's constant has been the subject of many revisions since 1930 when it was first introduced. *Hubble* used cepheid variables to calculate the distance of Andromeda and deduced a value for H of 530 km/sec/Mpc. Later on, *Baade* (1950) realised that the cepheids in Andromeda are 1.5 magnitudes brighter than it had previously been thought. The same was true for the few cepheids in the globular clusters. The implication was that Andromeda is at twice the distance that it was

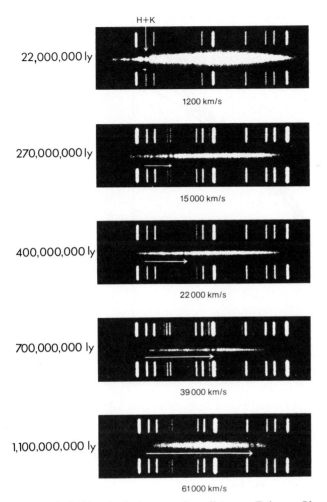

Fig. 3.6. Redshifts of galaxies at various distances (Palomar Observatory)

previously thought. After that, the distances of all galaxies had to be doubled. *Baade* deduced that $H = 260$ km/sec/Mpc.

Humason, Sandage and *Mayall* in 1956 studied the redshifts of 800 galaxies and deduced that $H = 180$ km/sec/Mpc. Their study has shown that Hubble's law applies only approximately to the distant galaxies, i.e. they realised that the velocity-distance relation is not exactly linear. Instead, there is some deceleration effect which becomes important when the distances become large. This deceleration is characterized by the parameter q which we shall examine in Sect. 6.4. In 1958 *Sandage* suggested that $H = 75$ km/sec/Mpc taking into account the revised absolute magnitudes of the brightest stars in galaxies.

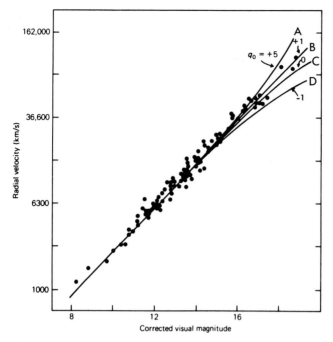

Fig. 3.7. Redshift as a function of the bolometric magnitude m of galaxies. The theoretical curves depend upon the value of the "deceleration parameter" q in (6.32): $A(q = 1)$, $B(q = 1/2)$, $C(q = 0)$ and $D(q = -1$, the value given by the steady state theory)

Table 3.1. Distances of various galaxies

Source	Distance in kpc or Mpc	Distance in light years
Centre of our Galaxy	8.5 kpc	28 000
Magellanic clouds	60 kpc	200 000
Andromeda galaxy	690 kpc	2 200 000
Edge of the local system	1000 kpc	3 300 000
M 51 (spiral galaxy)	3800 kpc	12 000 000
Centaurus A (radio galaxy)	4400 kpc	14 000 000
Virgo cluster	20 Mpc	66 000 000
Coma cluster ($z \simeq 0.02$)	40 Mpc	130 000 000
Hydra cluster ($z \simeq 0.2$)	1000 Mpc	3 300 000 000
Quasar (with $z = 2$)	5000 Mpc	16 000 000 000

Later on, *Sandage* and *Tammann* deduced an even lower value for the Hubble constant, $H = 50 \pm 7$ km/sec/Mpc, basing their research on studies of several galaxies. *Van den Bergh,* however, in 1965 suggested the value $H = 100$ km/sec/Mpc by taking into account all available data. *De Vaucouleurs* gave initially the value $H = 75$ km/sec/Mpc but revised it recently

(1982) to 100 ± 10 km/sec/Mpc. Thus the Hubble constant is believed to be between 50 and 100 km/sec/Mpc.

One can construct from the observations a diagram where the redshift of a galaxy is plotted against its bolometric[1] magnitude. Figure 3.7 shows such a graph with the dots corresponding to observational data and the curves A, B, C and D representing the theoretical predictions of the various models of the Universe (Sect. 6.4).

Table 3.1 gives a distance scale for various galaxies and clusters of galaxies.

---

[1] The bolometric magnitude of a star or a galaxy includes all types of radiation we receive from that particular object. The apparent and absolute bolometric magnitudes of an object, m and M respectively, and its distance from us in pc, r, are related by Eq. (2.2), which can be written:

$\log r = 0.2\, m + \text{constant},$

assuming M constant. Thus if $\log z$ is given by a similar expression (Fig. 3.7)

$\log z = 0.2\, m + \text{constant}$

we derive $z = \text{const} \times r$, which is Hubble's law for the expansion of the Universe (2.4).

Part II

Theory

4. Introduction to the Study of the Universe

4.1 Olbers' Paradox

The first observation concerning the whole Universe can be made with only one glance at the night sky, without any telescope and with limited theoretical knowledge. At the beginning of the last century, the german astronomer *Olbers* posed the following question: "Why is the night sky dark?" The answer to this question appears simple at first sight: "because there are not enough stars in the sky and therefore there are gaps between the stars". This answer is, however, wrong. There are enough stars to fill the sky and actually their total light "should" be sufficiently intense to set fire to the Earth. Let us examine a simple proof of this claim.

If every star radiates energy E_0 per unit time and if the Universe is homogeneous and infinite, the total energy of the light from all stars is, of course, infinite as well. To find the intensity of the light incident on the surface of the Earth we can make the following calculation. If a star is at a distance r its light is spread over the surface of a sphere with radius r, so that the intensity of radiation received on the surface of the Earth is

$$\Phi = \frac{E_0}{4\pi r^2} \, \pi R^2, \tag{4.1}$$

where R is the radius of the Earth. The number of stars in a spherical shell centred on the Earth, with radius r and thickness Δr is $4\pi r^2 \varrho \Delta r$, where ϱ is the mean number density of stars in space. Therefore the total light received by the Earth from this shell is

$$\Delta\Phi = \Phi 4\pi r^2 \Delta r \varrho = E_0 \varrho \pi R^2 \Delta r. \tag{4.2}$$

The number of such shells, however, is infinite, that is the total light is infinite.

This calculation does not take into account the fact that the light from the distant stars may be partly obstructed by the presence of intervening stars. A more accurate calculation (Sect. 4.2) shows that the light received theoretically per unit area of the Earth's surface is equal to the light emitted per unit area from the surface of a star, e.g. the Sun.

This result, however, contradicts our experience. The sky is dark, although our calculations show that it should be as bright (and as hot) as the surface of the Sun. This contradiction between observation and theory is called "Olbers' paradox".

Olbers' paradox forces us to reexamine the assumptions made in deriving this wrong theoretical result. The assumptions were the following:

a) The Universe is homogeneous (that is, the density of stars ϱ as well as the energy they emit per unit time are quantities constant in space).
b) Space is Euclidean, therefore the Universe is infinite.
c) The average value of the energy emitted, E_0, by the stars is constant in time.
d) The Universe is static.

If we drop the first assumption, (i.e. if we accept that the density ϱ becomes zero after a certain radius) it is easy to show that the light from the stars is finite. Nevertheless, the homogeneity of the Universe is one of its basic characteristics and we would not like to drop it except as a last resort.

The second assumption is not essential. That is, if we accept that the Universe is described by a spherical or a hyperbolic geometry (Sect. 6.2) our results will not change. In the case of hyperbolic geometry the Universe is infinite, as in Euclidean geometry. In the case of spherical geometry, the Universe is finite but without boundaries (just as the surface of a sphere is a two dimensional space which is finite without boundaries). In this second case, the light from a galaxy moves for ever in the closed Universe and passes an infinite number of times through the position of the observer. If the distance to a galaxy is τ light years and the time it takes the light to go once round the Universe is T, we receive not only the light emitted by the galaxy τ years ago, but also the light emitted $\tau + T$ years ago, and $\tau + 2T$ years ago, and so on. Thus the total amount of light we receive from all the galaxies is of the same order of magnitude as the amount of light we receive in a Euclidean Universe. Therefore if we drop the second assumption, the result is not going to change.

The third assumption, that the energy, E_0, emitted by the stars per unit time does not change, is obviously wrong. Stars do not live for ever. Every star is born, evolves and dies in a finite number of years. Even if we assume that the dead stars contribute to the creation of new stars, the life of the Universe is still not infinite. It must be less than the time required for all the matter in the Universe to be converted into energy, that is time $T_{max} = M c^2/E_0$, where M is the average mass of each star. This means that we receive light from stars only as far as T_{max} light years from us, therefore the total light of the Universe is finite.

The last remark above implies that the Universe is not eternal. Some people, in order to avoid this conclusion, assumed that the Universe is continuously being created, that is, new matter is continuously being created out of nothing (Sect. 7.3). If this is the case, the Universe would be eternal in the past and the future, and T_{max} would be infinite. This leads us to examine our last assumption, that the Universe is static.

Indeed, if the Universe is not static, but is continuously expanding, the light received from the most distant stars is much less than if the same stars were not been moving. Thus the total light is small, even though the Universe may be infinite and eternal.

From studying Olbers' paradox, therefore, we come to the conclusion that the Universe is either of finite age (if new matter is not continuously being created) or expanding.

If the Universe is expanding and no new matter is being created, then there must have been an origin of the expansion. If, however, new matter is being continuously created, the age of the Universe may be infinite, although every specific object, a star or a galaxy, in it has a finite age.

Olbers' paradox shows how many interesting syllogisms we can make by starting from one apparently insignificant observation, i.e. that the sky is dark at night. Of course, the study of astronomy gives us much more information with cosmological significance, as we shall see later on.

4.2 Calculation of the Energy Which Comes from all the Stars

Let us assume that all the stars have the same radius R_0 and the same temperature T. Then, according to the Stefan-Boltzmann law, they emit energy ε per unit time per unit surface area, given by $\varepsilon = \sigma T^4$, where σ is some universal constant. Therefore the total energy they emit per unit time is $E_0 = 4 \pi R_0^2 \varepsilon$.

Let us assume next that the energy which reaches the surface of a sphere with radius r, centred on the star, is E. E is less than E_0 because part of the light from the central star has been absorbed by other stars inside the sphere. This energy is spread all over the surface of the sphere, which is $4 \pi r^2$. Let us consider that a shell which surrounds this sphere has thickness Δr. Then the number of stars of this shell is $4 \pi r^2 \Delta r \varrho$. Every star inside this shell cuts down a beam of light from the central star equal to πR_0^2, that is the fraction of the total light absorbed by the stars in the shell is

$$\frac{\Delta E}{E} = - 4 \pi r^2 \Delta r \varrho \, \frac{\pi R_0^2}{4 \pi r^2} = - \varrho \pi R_0^2 \Delta r, \tag{4.3}$$

where ΔE is considered negative since the light from the central star is reduced. If we integrate the above equation considering that ΔE and Δr are differentials, we find that the total light reaching a distance r from a given star is

$$E = E_0 \exp(-\varrho \pi R_0^2 r). \tag{4.4}$$

The energy received by the Earth is $\pi R^2 E/(4\pi r^2)$ where R is the radius of the Earth. Further, the number of stars inside a shell centred on the Earth and having thickness Δr is $4\pi r^2 \Delta r \varrho$. Therefore, the energy received at the Earth from all the stars in this shell is

$$4\pi r^2 \Delta r \varrho \frac{\pi R^2 E}{4\pi r^2} = \pi R^2 \varrho \cdot 4\pi R_0^2 \cdot \varepsilon \exp(-\varrho \pi R_0^2 r) \cdot \Delta r. \tag{4.5}$$

Given that this energy is evenly spread all over the surface of the Earth, which is $4\pi R^2$, the energy received by the Earth per unit time and unit surface area is given by

$$\Delta\phi = \varrho \pi R_0^2 \cdot \varepsilon \exp(-\varrho \pi R_0^2 r) \cdot \Delta r. \tag{4.6}$$

If we integrate this equation assuming that r varies from zero to infinity, we eventually find that

$$\phi = \varepsilon. \tag{4.7}$$

That is, the energy received per unit surface area of the Earth, is equal to the energy emitted by an average star per unit surface area of the star. This is the theoretical calculation that leads to Olbers' paradox.

5. General Theory of Relativity

5.1 The Concept of Curved Spacetime

The general theory of relativity of *Einstein* gave the greatest boost to the theoretical study of the Universe.

The special theory of relativity (*Einstein* 1905) constitutes today an integral part of classical physics. However, in spite of its great success, this theory left some basic questions unanswered. The principle of special relativity tells us that "the laws of nature are the same in all inertial frames". An inertial frame is a co-ordinate system in which the principle of inertia applies, i.e. when no forces act on a particle, it either remains stationary or moves on a straight line with constant velocity. Any co-ordinate system which is in uniform motion with respect to an inertial frame is also an inertial frame. But why should the laws of nature be the same only in inertial frames? What happens when a frame is subject to acceleration, i.e. when the principle of inertia does not hold in it? When the motion of a particle is not uniform, we say that some forces are acting on it. It is possible, however, that in fact no forces are acting on the particle, but the co-ordinate system from which we view it is non-inertial[1]. *Einstein* realised immediately the weakness of the special theory of relativity and proceeded, in 1916, to propose the general theory of relativity, which generalises the principle of relativity by stating that "the laws of nature are the same in two frames which move in any possible way with respect to each other".

As a mathematical tool for the application of this principle *Einstein* used tensor calculus (Sect. 5.2). Tensors are generalisations of vectors, but they have more components than vectors. Their major characteristic is that the equality of two tensors does not depend upon the co-ordinate system. If two tensors are equal in one co-ordinate system they remain equal in any system which moves in any possible way with respect to the first one. Therefore, if a physical law is expressed as an equality of two

---

[1] For example, motion along a straight line may appear curved to a rotating observer. In this way, however, we cannot say with certainty whether the system is inertial or not. This difficulty is not only found in the theory of relativity, but in classical mechanics as well. We usually avoid it by defining an approximately inertial frame which is "centred on the Sun and its three axes are directed towards three distant galaxies".

tensors, it is independent of the co-ordinate system. The basic law of general relativity is exactly of this form. It is expressed by the field equations of *Einstein* which relate the curvature of spacetime with the distribution of matter and energy.

Spacetime is a system of four co-ordinates, one of which represents time and three represent space. The form of spacetime is determined by the so-called "metric tensor". If two neighbouring points in spacetime are $\Sigma(x_0, x_1, x_2, x_3)$ and $\Sigma'(x_0 + dx_0, x_1 + dx_1, x_2 + dx_2, x_3 + dx_3)$ the distance between them is called ds, and is given by the formula

$$ds^2 = \sum_{\mu=0}^{3} \sum_{\nu=0}^{3} g_{\mu\nu} dx_\mu dx_\nu, \tag{5.1}$$

where the quantities $g_{\mu\nu}$ are functions of x_0, x_1, x_2, x_3, and the summations Σ refer to μ and ν varying from 0 to 3. The quantities $g_{\mu\nu}$ constitute the metric tensor, which is symmetric, i.e. $g_{\mu\nu} = g_{\nu\mu}$. The expression (5.1) for the square of the elementary distance, ds, is called the "line element". A particularly simple case of the metric tensor is the tensor of special relativity, or Minkowski tensor, which has $g_{00} = 1$, $g_{11} = g_{22} = g_{33} = -1$, while the quantities $g_{\mu\nu}$ with μ different from ν are zero. That is, the line element of Minkowski spacetime is

$$ds^2 = dx_0^2 - dx_1^2 - dx_2^2 - dx_3^2. \tag{5.2}$$

In this equation x_1, x_2 and x_3 are the three co-ordinates of ordinary space, and $x_0 = ct$ where t is the time and c is the speed of light. In this case we say that space is pseudo-Euclidean. This is because, if we define three "imaginary" directions of space by setting $x_1 = ix_1'$, $x_2 = ix_2'$ and $x_3 = ix_3'$ where $i = \sqrt{-1}$, we have

$$ds^2 = dx_0^2 + dx_0'^2 + dx_2'^2 + dx_3'^2, \tag{5.3}$$

which is the four-dimensional generalisation of the distance formula in Euclidean space (Pythagoras' theorem). The Minkowski spacetime is also called "flat" or "not curved" spacetime. It can easily be shown that in such a spacetime the motions of the particles and of the light are linear and uniform. In general, any spacetime, the line element of which can take the form (5.2) by a change of co-ordinates, is flat. The line element in polar co-ordinates is such an example:

$$ds^2 = c^2 dt^2 - dr^2 - r^2(d\theta^2 + \sin^2\theta \, d\phi^2). \tag{5.4}$$

There are, however, spaces which are curved and they cannot be transformed into flat ones. Such a curved space of two dimensions is the surface of a sphere. For example, two neighbouring points on the surface of the Earth are at a distance from each other:

$$ds^2 = d\phi^2 + \cos^2 \phi \, d\lambda^2 \qquad (5.5)$$

where λ and ϕ measure the geographical longitude and latitude, respectively. In this case there is no transformation which will make ds^2 to look like the sum of two squares $dx_1'^2 + dx_2'^2$.[2]

In general, every four-dimensional space is characterised by a fourth-order tensor, called the Riemann tensor $R_{\kappa\lambda\mu\nu}$. When $R_{\kappa\lambda\mu\nu}$ is different from zero, spacetime is curved and when $R_{\kappa\lambda\mu\nu}$ is equal to zero, it is flat.

The general theory of relativity deals, in general, with curved spacetimes. In such a spacetime the motions of particles, and of light, are curved. However, these curves have a common characteristic with the straight lines. Just as straight lines are the shortest paths connecting two points, the motions in curved spacetimes are the shortest curves between two points. Such curves are called geodesics. For example, on the surface of a sphere one can only draw curves and not straight lines. From all the curves which connect two points, the shortest is the arc of a great circle. Therefore, the geodesics on the surface of a sphere are the arcs of great circles.

It is well known that light travels along the fastest route between two points. In a medium with constant refractive index, i.e. in a vacuum or in air of constant density, the light path is a straight line. If, however, the refractive index varies, the light path is a curve. This is the effect of refraction which makes the stars appear to be higher in the sky, above the horizon, than they really are. In this case, the fact that the light does not move uniformly on straight lines is not attributed to any force, but to the change in its velocity, which in turn is due to the change in the air density. Something similar happens in curved spacetime. The light follows geodesic curves. We say that the light does not move uniformly along straight lines, not because it is subject to some force, but because the spacetime is curved. The concept of force, therefore, has been replaced by the geometric concept of the curvature of spacetime.

A relevant example is the frictionless motion of a golf ball on an uneven terrain. The ball follows the dips and bumps of the ground, up and down according to the curvature. We may say that this motion is determined by the way the ball was initially thrown, and the force of gravity. Alternatively, we may say that the ball always follows the curvature of the

---

[2] More accurately, there is no transformation which gives x_1' and x_2' as functions of λ and ϕ globally. We note this because close to a *given point* $\lambda = \lambda_0$ and $\phi = \phi_0$, there is always a transformation which defines *locally* a flat spacetime (in our case $x_1' = \phi$ and $x_2' = \lambda \cos \phi_0$). That is, spacetime can be assumed to be flat locally. In the same way a curved surface can be approximated locally by its tangent plane at a point.

ground, that is why its motion is not simply along a straight line. This second way of viewing things corresponds to the picture of general relativity.

5.2 Tensors and the General Theory of Relativity

It is impossible to include in this book a detailed analysis of the general theory of relativity. The relevant references on the subject are included in the bibliography at the end of the book. We are simply going to give here some definitions and make some general remarks.

The metric tensor $g_{\mu\nu}$ in the four-dimensional spacetime consists of 16 components, i.e. 16 functions of the four co-ordinates x_0, x_1, x_2 and x_3.

Because of the symmetry $g_{\mu\nu} = g_{\nu\mu}$, only 10 of them are independent. If the functions $g_{\mu\nu}$ are to be components of a tensor, the following condition must hold: Suppose we change our co-ordinate system in any conceivable way, so that the co-ordinates x_0, x_1, x_2 and x_3 are any functions of the four new co-ordinates x'_0, x'_1, x'_2 and x'_3. Then, the components $g'_{\kappa\lambda}$ of the tensor in the new co-ordinate system must be given by

$$g'_{\kappa\lambda} = \sum_{\mu=0}^{3} \sum_{\nu=0}^{3} \frac{\partial x_\mu}{\partial x'_\kappa} \frac{\partial x_\nu}{\partial x'_\lambda} g_{\mu\nu} \tag{5.6}$$

where $\partial x_\mu / \partial x'_\kappa$ is the partial derivative of x_μ with respect to x'_κ.

The above formula is a double sum because the tensor has two indices. For a tensor with one index the sum is single. These tensors are of "second order" or "first order" accordingly. There are also tensors of "zeroth order" which do not have any indices and are scalars, and tensors of higher order derived in a similar way; the number of summations in (5.6) is equal to the order of the tensor. Further, the number of terms in each summation is equal to the dimension of the space where the tensor is defined.

In a Euclidean space of *three dimensions* the distance of two neighbouring points (x_1, x_2, x_3) and $(x_1 + dx_1, x_2 + dx_2, x_3 + dx_3)$ is given by the formula:

$$ds^2 = dx_1^2 + dx_2^2 + dx_3^2. \tag{5.7}$$

The metric tensor in this case is $g_{11} = g_{22} = g_{33} = 1$ and $g_{\mu\nu} = 0$ for $\mu \neq \nu$. If we change into polar co-ordinates defined by

$$x_1 = r \sin\theta \cos\phi, \qquad x_2 = r \sin\theta \sin\phi, \qquad x_3 = r \cos\theta, \tag{5.8}$$

formulae (5.6) can be used to calculate the metric tensor in the new co-ordinates, provided we call x'_1, x'_2 and x'_3, r, θ and ϕ respectively. We find $g'_{11} = 1$, $g'_{22} = r^2$, $g'_{33} = r^2 \sin^2 \theta$ and $g_{\kappa\lambda} = 0$ for $\kappa \neq \lambda$. Thus, the line element in polar co-ordinates is

$$ds^2 = dr^2 + r^2(d\theta^2 + \sin^2 \theta \, d\phi^2). \tag{5.9}$$

Having defined $g_{\mu\nu}$ one can proceed to define the Einstein tensor $G_{\mu\nu}$ which is a function of $g_{\mu\nu}$ and of their first and second derivatives with respect to the variables x_κ.

The $G_{\mu\nu}$ tensor appears in Einstein's field equations (Sect. 5.3) which relate it to the energy-momentum tensor $T_{\mu\nu}$. The latter is defined by the formula:

$$T_{\mu\nu} = (\varrho c^2 + p) \, u_\mu u_\nu + p g_{\mu\nu}, \tag{5.10}$$

where u_κ are the components of the velocity, ϱ is the matter-energy density and p is the pressure. (It is assumed that matter and energy are equivalent, according to the formula $E = mc^2$.) Therefore, the energy-momentum tensor depends upon the density and the pressure of the matter-energy, as well as the motion of the matter-energy in space.

5.3 Einstein's Field Equations

Einstein's theory not only tells us that spacetime is curved, but it also specifies *how much* its curvature is. More specifically, it gives a set of equations which relate the curvature of spacetime with the distribution of matter-energy in space. These equations are called field equations and have the form:

$$G_{\mu\nu} = -\kappa T_{\mu\nu}, \tag{5.11}$$

where $\kappa = 8\pi G/c^4$ and G is the constant of gravity. The "Einstein tensor" $G_{\mu\nu}$ depends upon the functions $g_{\mu\nu}$ and their first and second derivatives, while the tensor $T_{\mu\nu}$ is the "energy-momentum tensor" and depends upon the distribution of the energy and matter in space (Sect. 5.2).

Equation (5.11) means that the curvature of spacetime is due to the distribution of the mass-energy in space. If we solve the above equations we find the metric tensor $g_{\mu\nu}$, which is due to the tensor $T_{\mu\nu}$. Equation (5.11) corresponds to Poisson's equation, that relates the potential to the density of matter. In that case, the potential specifies the forces which act on a certain particle. That is why the components of the metric tensor $g_{\mu\nu}$ are sometimes called "potentials". Einstein's field equations constitute a special application of "Mach's principle", according to which the inertial properties of matter are due to the distribution of the matter in the rest of the Universe.

Mach's basic observation was that the velocity and the acceleration of a particle would be meaningless if the particle were alone in the Universe. We may talk only of accelerations with respect to other bodies, just as we talk of velocities with respect to other bodies. The concept of relative velocity led to special relativity. The concept of relative acceleration is the major ingredient of Mach's principle, which led *Einstein* to develop his general theory of relativity. Let us take, as an example, the rotation of the Earth about its axis. The Earth rotates, not with respect to any absolute space, but with respect to the distant stars in the Universe. If the Earth were completely covered by thick clouds, we would still be able to find its rotation by using Foucault's pendulum. A pendulum at the North Pole of the Earth turns its plane gradually around, with respect to the Earth, since its plane is kept fixed with respect to the distant stars. If no other stars existed in the Universe, apart from the Earth, according to Mach's principle the plane of the pendulum would remain constant with respect to the Earth. Therefore, in some way the distant matter in the Universe has consequences which concern the behaviour of matter around us. *Einstein* tried to incorporate this principle into his general theory of relativity.

If the tensor $T_{\mu\nu}$ is zero everywhere, that is, if there is no matter in the Universe, one solution of the field equations (5.11) is the Minkowski "flat" spacetime.

Another relatively simple solution of the field equations concerns a spherical body in an empty space. If we consider the Sun as a spherical body and assume that space is empty around it, Eq. (5.11) may give us the curvature of spacetime in that space. The metric tensor around the Sun is called the "Schwarzschild metric " (1916) and corresponds to a line element which, in polar co-ordinates, is

$$ds^2 = c^2\left(1 - \frac{2GM}{c^2r}\right)dt^2 - \frac{dr^2}{1 - \dfrac{2GM}{c^2r}} - r^2(d\theta^2 + \sin^2\theta\, d\phi^2). \quad (5.12)$$

5.4 Experimental Verification of the General Theory of Relativity

We observe that if the mass M of the body which gives rise to the gravitational field described by (5.12) is very small, the spacetime is approximately the same as the Minkowski spacetime (5.4). As a next step in (5.12)

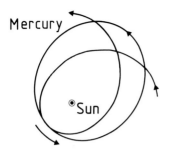
Mercury

Fig. 5.1. Mercury's perihelion shift

Sun

we omit terms of order $1/c^2$ assuming that $2GM/c^2 \ll r$, and obtain

$$ds^2 = \left(c^2 - \frac{2GM}{r}\right) dt^2 - dr^2 - r^2(d\theta^2 + \sin^2\theta \, d\phi^2). \tag{5.13}$$

This is the "Newtonian approximation" because it reproduces the results of Newton's theory for the motions of the planets. That is, in this approximation the orbits of the planets are the ellipses given by Newton's law of gravitational attraction.

If, however, we consider the geodesics in Schwarzschild's spacetime, defined by (5.12), we see that the orbits diverge from those obtained by the application of Newton's law. For example, the orbits of the planets are ellipses with their major axes gradually shifting around (Fig. 5.1). This effect is more prominent in the case of Mercury which is the closest planet to the Sun. One of the major verifications of general relativity has been the very accurate prediction of the observed orbit of Mercury. Indeed, the theory predicts that the perihelion of Mercury's orbit shifts around by 43.03″ per century, while observations yield 43.11″ ± 0.45″ per century. Similar effects, but much weaker, have been observed in the orbits of Venus and the Earth.

A much stronger effect, of the same nature, was observed in 1979 with the discovery of the double pulsar PSR 1913 + 16[3]. In this case the perihelion shift is 4.2° per year, which is enormous in comparison to the shift in Mercury's orbit. In general, this double pulsar has given us many opportunities to verify the predictions of general relativity.

Another prediction of general relativity is that light rays passing close to the Sun bend (Fig. 5.2). This phenomenon is indeed observed during a total eclipse when stars close to the Sun become visible. These stars are observed to have shifted away from the positions they have when the Sun is in another part of the sky. There is very good agreement between the shift predicted by the theory, 1.75″, and that observed, 1.73″ ± 0.05″.

---

[3] The relativistic effects in double pulsars are important, because of their enormous densities (up to 10^{14} g/cm^3) and minimal sizes, that allow them to come very close to each other.

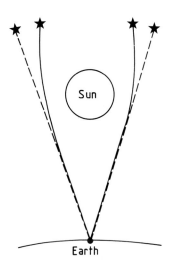

Fig. 5.2. Deflection of light by the Sun (*continuous lines*). We see the stars in the direction of the dashed lines

A third prediction of the theory is that the light from stars with strong gravitational fields is more redshifted then the light from similar sources with weak gravitational fields. Indeed, the light from the Sun is redshifted in comparison with the light from corresponding terrestial sources. The relative redshift is

$$\Delta = \frac{GM_\odot}{R_\odot c^2}, \tag{5.14}$$

where $M_\odot$ and $R_\odot$ are the mass and the radius of the Sun respectively. The most recent observations give values of Δ which are 1.05 ± 0.05 times the theoretically predicted value. This effect is much more marked in white dwarfs where the gravitational field is much stronger.

The same phenomenon, but much weaker, may be observed on the Earth. It is known that the gravitational field close to the surface of the Earth is stronger than at higher altitudes. It is expected, therefore, that the spectral lines of a certain atom will be slightly redshifted when the atom is on the surface of the Earth, in comparison with the lines of the same atom when it is at a certain height above the Earth. In spite of the fact that the expected redshift is very small, this effect has been observed by *Pound* and *Rebka* in 1959. *Pound* and *Rebka* managed to measure the relative redshift between two sources at 22.5 metres difference in height, by using very sensitive instruments. They expected to find a relative difference in the wavelength of the lines $\Delta\lambda/\lambda = 2.5 \times 10^{-15}$ and they actually observed such a difference with 1% accuracy.

The three effects we have just described constitute the classical tests of relativity, which have established the theory. In recent years, more tests have been devised. The most important of them concerns the delay in

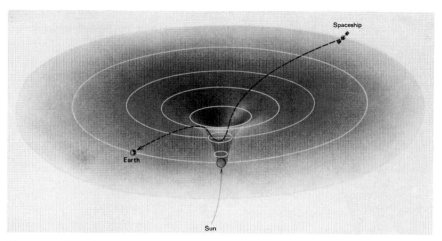

Fig. 5.3. A signal from a spaceship to the Earth arrives with delay because of the curvature of space within the gravitational influence of the Sun

receiving back the reflected radar signals sent to other planets, or to spaceships (*Shapiro* 1964) (Fig. 5.3). The observed value of this delay is 1.000 ± 0.002 times the theoretically predicted one. This method can be improved in the future to reach accuracies of 0.1%.

The above tests have to be passed by other theories of gravitation, which are put forward from time to time hoping to replace general relativity. For example, the Newtonian theory based on formula (5.13) predicts an angle of divergence for the light rays passing close to the Sun, which is half of that predicted by general relativity. Further, this theory cannot explain the shifting of Mercury's perihelion. That is why it is rejected. The same classical tests significantly restrict the possible application of the Brans-Dicke theory (Sect. 7.4) and other theories, and even allow us to reject some of them entirely.

The tests we described, and other similar ones, refer to *small deviations* of general relativity from Newtonian theory. The presumption is that the gravitational fields are weak and the velocities of the particles are small in comparison with the speed of light. In such cases the various quantities are usually expanded into power series of $1/c^2$. For example, the term $dr^2/[1 - 2GM/(c^2 r)]$ of *Schwarzschild's* line element, can be written as

$$dr^2 \left(1 + \frac{2GM}{c^2 r} + \frac{4G^2 M^2}{c^4 r^2} + \cdots \right).$$

We recognise the first term as that of the Newtonian theory. The second term $dr^2 \cdot 2GM/(c^2 r)$ is the first post-Newtonian term and so on. The theory of the post-Newtonian approximations allows us to check the accuracy of the various tests of general relativity, as well as to compare general relativity with other theories.

5.5 Gravitational Waves

A particularly interesting set of solutions of the field equations are those which represent *gravitational waves*. If we assume that the spacetime is only slightly different from the flat Minkowski spacetime and solve the linearised field equations, we find a solution which represents gravitational waves. The gravitational waves are similar to the electromagnetic waves, i.e. they propagate with the speed of light.

The question of the origin of these waves then arises. It is easy to accept that gravitational interactions propagate with the speed of light rather than instantly, with infinite velocity. Every change, therefore, in a mass which will cause a change in the gravitational field, will propagate with the speed of light. For example, if two stars collide to form a new spherical body, the new gravitational field will be a point mass field instead of a field created by two separate sources. This change will not be felt instantly at distance r, but after a time $t = r/c$, i.e. the time required for the gravitational change to propagate to a distance r. Every such change can be considered as a gravitational wave which propagates with velocity c. This is particularly evident in the case of a periodic phenomenon, e.g. the rotation of a star or the motion of a double star. The gravitational wave in that case has the same frequency as the rotation, and is very similar to an electromagnetic wave of certain frequency. On the other hand, we must remark that a spherical body which pulsates does not change its gravitational field outside its surface and therefore, it does not produce gravitational waves. Thus, the spherical collapse of a star does not produce gravitational waves. However, the collapse of a spheroidal star, or the fall of a body into a black hole (Sect. 5.6), does produce gravitational waves.

The gravitational waves transfer energy. Therefore, this energy must be subtracted from the body, or the system of bodies, which have emitted them. Every change in the source due to the emission of gravitational waves is called "radiation reaction".

For example, the energy and angular momentum of a rotating star which emits gravitational waves changes. In this way we explain the continuous decrease in the rotation of pulsars (neutron stars, Sect. 1.10). The pulsars rotate very rapidly, and at the same time they send us flashes of light with period equal to their rotational period (from 0.0015 sec to 3 sec approximately). Obviously neutron stars have strong gravitational fields and therefore, if they are not spherical they must emit strongly gravitational waves. As a consequence, these stars must continuously lose energy and angular momentum, and thus be decelerated, as observed. Their period gradually changes by 10^{-5} sec per year. The strongest verification so far that gravitational waves exist comes from the double pulsar PSR 1913 + 16. The change in the period of this system confirms not only qualitatively, but also quantitatively, Einstein's formula for the gravitational radiation:

$$I = \frac{32\,m^5\,G^4}{5\,d^5\,c^5},$$

(5.15)

where I is the power of the radiation, m is the mass of each star and d the distance between the two stars.

Thorne (1969) worked out theoretically the "reaction of the gravitational radiation" for the case of neutron stars. The general problem of the emission of gravitational radiation by any source was solved in 1970 by *Chandrasekhar* and his collaborators with the method of post-Newtonian approximations. The amount of energy radiated away by a pulsar is very small, according to formula (5.15). It is proportional to the fifth power of the mass and inversely proportional to the fifth power of the distance and of the speed of light. The radiation we expect to receive on Earth is too small to be measured. It is possible, however, to measure the radiation we receive from other, stronger sources. Such a source is the centre of our Galaxy which, according to many people, is an enormous black hole emitting gravitational waves as it continually swallows stars.

Many scientists have tried so far to detect experimentally gravitational waves received from space. *Weber* was the first to make such an experiment in 1960. He installed two similar cylinders, each weighing 1.5 tonnes, one in the university of Maryland, and one in the Argonne National Laboratory in Chicago, at a distance of 1000 km. He observed, with great accuracy, the various sudden vibrations of the cylinders. He particularly noted the simultaneous vibration of the two cylinders. Such vibrations could not be due to local causes, as the two cylinders were 1000 km apart. It is possible that they were due to gravitational waves which came from space, perhaps from the centre of our Galaxy. If *Weber*'s results are correct, the intensity of the gravitational waves emitted by the centre of our Galaxy is indeed enormous. This has made many scientists sceptical, so many scientific teams started similar experiments, more accurate than *Weber*'s. These experiments did not confirm his results. Thus, today it is considered rather unlikely that the simultaneous vibrations detected by *Weber* were due to gravitational waves. However, no other better explanation has been proposed. An attempt has been made to install a gravitational wave detector on the Moon, but it failed. If it had succeeded, we would have had a better chance of detecting such waves, due to the large distance between the Moon and the Earth. It is hoped, however, that due to the efforts of many scientists, gravitational waves will be detected in the near future.

The gravitational waves we receive from particular sources are very weak. However, the gravitational waves generated in the very early Universe are expected to be much stronger, and they must constitute an important ingredient of the Universe. If we ever succeed in detecting these waves we shall have information about the condition of the Universe during the first thousandth of the first second of its life. Unfortunately, such an observation is considered entirely unlikely since these waves now correspond to a radiation of 1 K (i.e. only one degree above absolute zero).

5.6 Black Holes

Black holes are the most extreme predictions of general relativity. When the radius of a star becomes less than a certain limit, called the Schwarzschild radius

$$r_s = \frac{2GM}{c^2}, \tag{5.16}$$

neither light nor particles can escape to reach the Earth.

A simple explanation of this phenomenon is the following: We know from classical mechanics that the escape velocity of a particle from the surface of a star is

$$v_\infty = \sqrt{2GM/r}. \tag{5.17}$$

If the velocity of a particle is greater than v_∞, the particle moves to an infinite distance (neglecting, of course, the influence of the neighbouring stars). If, however, the velocity is less than v_∞, the particle cannot go to infinity and falls back onto the star.

When the escape velocity is equal to the speed of light, (5.17) defines the Schwarzschild radius. We may say, therefore, that if the radius of a star is very small, its gravitational field is sufficiently strong for the escape velocity to equal the speed of light. The Schwarzschild radius for the Sun is 3 km, and for the Earth 0.9 cm. That is, if the radius of the Sun were less than 3 km, its light could not be emitted. The same would hold for the Earth if its radius were less than 0.9 cm.

The above explanation does not interpret all the phenomena related to black holes. The full theory of black holes is based on general relativity (Sect. 5.7).

Black holes are very important in astronomy, and particularly in cosmology, for the following reasons:

a) Black holes are the final product of the collapse of stars more massive than three solar masses. Therefore, many stars have already become, or will become, black holes.

b) Mini black holes might have been formed in the early Universe out of irregularities in the gravitational field. We are going to discuss the possible existence of such black holes in Sect. 8.7.

c) If the existence of black holes is established beyond doubt, we will have one of the strongest verifications of general relativity. Therefore, our trust in this theory, so important in cosmology, will be justified. This is particularly relevant to the initial Big Bang and the final collapse of the Universe, because the conditions are then similar to those of a black hole.

d) The formation of black holes is very important for the future evolution of the Universe (Sect. 9.2).

For all of the above reasons it is worthwhile proceeding to a more detailed description of black holes.

5.7 Formulae Related to Black Holes

Schwarzschild's line element can be written as follows:

$$ds^2 = c^2 \left(1 - \frac{r_s}{r}\right) dt^2 - \frac{dr^2}{1 - r_s/r} - r^2(d\theta^2 + \sin^2\theta \, d\phi^2) \tag{5.18}$$

where r_s is the Schwarzschild radius. When r becomes equal to r_s, the coefficient of dr^2 becomes infinite. Further, while outside the black hole the coefficient of dt^2 is positive and that of dr^2 negative, inside the black hole the reverse is true. The implication is that inside the black hole the r co-ordinate represents time while the t co-ordinate represents space. In fact, the distinction between a timelike and a spacelike elementary length ds is given by the sign of ds^2; ds^2 is positive for a timelike length and negative for a spacelike length (i.e. in the latter case ds is imaginary). Thus, outside the black hole we have a timelike length if we put $dr = d\theta = d\phi = 0$ in which case

$$ds^2 = c^2 \left(1 - \frac{r_s}{r}\right) dt^2 > 0; \tag{5.19}$$

inside the black hole we have a timelike length if we set $dt = d\theta = d\phi = 0$:

$$ds^2 = -\frac{dr^2}{1 - r_s/r} > 0. \tag{5.20}$$

Therefore, inside the black hole the r co-ordinate represents time and not space.

A light ray follows a path such that $ds^2 = 0$. If a light ray comes from a large distance straight to the black hole ($d\theta = d\phi = 0$) using formula (5.18) we obtain

$$dt = -\frac{dr}{c(1 - r_s/r)};$$
(5.21)

the negative sign is due to the fact that light moves towards the black hole, i.e. $dr < 0$.

To calculate the total time required for the light to reach the black hole, we integrate (5.21). In this way, we find that the light requires infinite time ($t \to \infty$) to reach the Schwarzschild radius r_s, if it starts from a distance $r_0 > r_s$. The same also holds for the motion of a particle.

A moving observer, however, measures his own time τ which is defined by

$$ds^2 = c^2 d\tau^2.$$
(5.22)

This is called "proper time". This time remains finite, i.e. the total time τ required for an observer to reach radius r_s, falling from $r_0 > r_s$, is finite.

Similarly, the proper time required for an observer to reach the singularity $r = 0$ starting from $r = r_s$ is also finite. In this way it can be shown that once an observer enters a black hole, he will inevitably reach the central singularity in finite time.

This, however, is not necessarily true for a rotating black hole (Sect. 5.10). There are cases when a particle can fall into a black hole and come out again. It will take infinite time t to come out again, but finite time τ. We therefore have the paradox that the observer will come out in "another world" after infinite time has passed in our world.

5.8 Schwarzschild Black Holes

Schwarzschild in 1916 derived formula (5.12) which describes the basic type of a black hole. Such a black hole is spherically symmetric. Schwarzschild's radius (5.16) defines a spherical surface, called "event horizon", which is the limit of the black hole. It is obvious from (5.12) that as r reduces and approaches r_s, Schwarzschild's spacetime becomes very different from Minkowski spacetime. The spacetime inside a black hole is even more different.

A particle (or a photon) which approaches the black hole coming from a large distance is attracted and its orbit diverges from a straight line (Fig. 5.4). As the particle approaches the black hole, this divergence in-

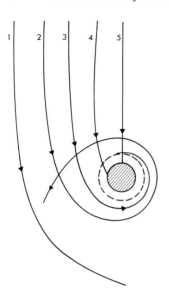

Fig. 5.4. Orbits close to a black hole

creases. It is possible for the particle to go around the black hole once or several times before it escapes to infinity (orbit 2 in Fig. 5.4). A limiting case for such an orbit is orbit 3 which goes around the black hole infinite times and approaches asymptotically the sphere with radius $3r_s/2$. In general, if a particle goes too close to a black hole it is trapped and it cannot escape to infinity. More specifically, if the particle travels straight to the centre of the black hole (orbit 5), it falls inside it and is lost forever. An observer, however, far from the black hole, will see such a particle taking infinite time to reach the Schwarzschild radius (Sect. 5.7). The same is true for orbits which do not deviate much from orbit 5.

That is, the distant observer will see particles which follow orbits like orbit 5 and 4, approaching the black hole with ever decreasing velocity, but he will never see them actually entering the black hole. If the falling particle happens to emit radiation, the distant observer will see its light redshifted since the source is under the influence of a strong gravitational field. The closer the particle goes to the black hole the more redshifted its light becomes. Thus the particle will eventually disappear, as the light from it will be redshifted into the far infrared, then to the radio waves and eventually to such long wavelengths that it will be impossible to detect.

The strange thing is that the proper time of an observer falling in with the particle remains finite. This is an extreme case of "time dilation" when the finite time interval of the moving observer is seen as infinite by the distant observer (as $r \to r_s$, $t \to \infty$, while τ remains finite, see Sect. 5.7). As the falling observer crosses the event horizon he does not feel anything special. He simply feels that the gravitational force increases. As he approaches the centre, however, the increase in the gravitational force be-

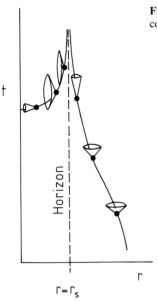

Fig. 5.5. Orbit of a particle entering a black hole. The light cone has been drawn at various positions

Horizon

$r = r_s$

comes enormous. The tidal forces increase, and are sufficient to tear him apart. No other force, electromagnetic or nuclear, can prevent this tearing. That is, not only is the observer torn apart but so too are his atoms and protons. (This fate may be avoided in a Kerr black hole; see Sect. 5.10.)

Finally, only a finite time after the observer enters the horizon, he reaches the centre where he ends up as a mathematical point.

Such a fate awaits any material object which may enter a black hole (Fig. 5.5). The centre of a black hole is a "mathematical singularity" where matter has infinite density. Its "event horizon" acts like a semitransparent membrane which allows anything to enter but prevents everything from coming out. Anything entering the black hole is doomed to end up in the central singularity.

Figure 5.5 shows the orbit of a particle as it approaches a black hole. The abscissa is the r axis while the ordinate is the t axis. At various points along the orbit the light cone is marked representing the light rays emitted from the particle. As the particle approaches the radius r_s, the time t tends to infinity. After it has entered the black hole, however, t decreases. The r co-ordinate heads towards smaller and smaller values, until it reaches $r = 0$. The light cone also aims towards the centre so that all the photons emitted by the particle also propagate towards the singularity $r = 0$.

One may wonder if it is possible for the reverse to happen, that is, the particles and light to follow an exactly opposite route and emerge out of a black hole.

At first sight Schwarzschild's line element can also describe the reverse process. In that case we speak of a "white hole". Particles and light come out of a white hole, and they can never go back into it. The proper time τ required for a particle to go from the singularity $r = 0$ to the horizon and beyond is finite. For an external observer, however, this time is infinite.

However, there is a basic problem concerning white holes. While black holes can be produced by physikal processes the origin of a white hole is inexplicable. The appearance of a white hole is not due to any cause, it is acausal. This is why, in spite of a small minority of authors who employ white holes to explain certain energetic phenomena in galaxies and quasars, most scientists reject the existence of white holes. Indeed, the "explanation" of a phenomenon based on an "inexplicable" event, like the creation of a white hole, is not really an explanation.

As we mentioned earlier, a particle falling inside a black hole always reaches the central singularity. It has been proposed that this singularity may be identified with the singularity inside a white hole, so that a particle reaching the singularity may come out again. If something like this can happen, it is possible for entirely unnatural situations to arise which violate causality. For example, a person might be able to meet himself at a different age and talk with, or even kill, his alter ego. At this point we stop really doing science and we are entering the world of science fiction. That is why mathematical solutions which lead to unnatural consequences, like the abolition of causality, or influence and possible alteration of one's own past, are rejected. Indeed, one of the basic restrictions we impose on solutions of Einstein's field equations is that they maintain causality.

Another interesting problem is the formation of black holes by the final collapse of a massive star.

The final collapse of a star occurs when all its nuclear energy fuel is exhausted. As is known, stars form out of concentrations of the interstellar medium. The gravitational attraction between the particles makes them collapse, resulting in an increase of density and temperature. When the temperature is high enough, nuclear reactions start in the centre of the star. They produce a lot of radiation which hinders any further collapse of the outer layers.

These nuclear reactions basically convert hydrogen into helium. Given that most of the star's mass is in hydrogen, this phase lasts many millions, or even billions, of years. When the hydrogen is exhausted, other nuclear reactions involving helium and other heavier elements start. These last for a relatively short time, a few dozens of millions of years. Finally, all nuclear reactions stop. Then there is not enough internal pressure to support the outer layers and there are three possibilities for the subsequent evolution of the star:

a) If the mass of the star is less than 1.4 $M_\odot$ (solar masses) the star becomes a white dwarf. *Chandrasekhar* has developed the theory of white dwarfs and the mass limit of 1.4 $M_\odot$ is called the Chandrasekhar limit. The

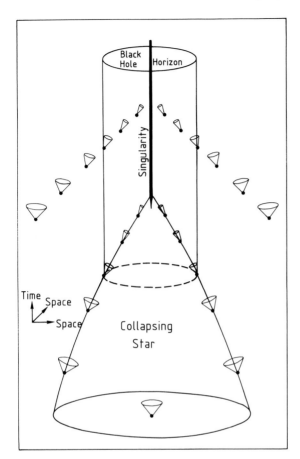

Fig. 5.6. A star collapsing to form a black hole

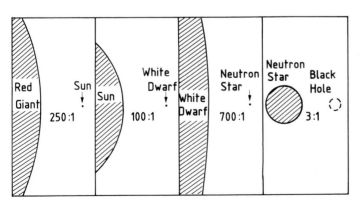

Fig. 5.7. Relative sizes of the Sun, a red giant, a white dwarf, a neutron star and the horizon of a black hole

density of a white dwarf is millions of times the density of water. The matter in this case is degenerate, i.e. Boltzmann's statistics and the usual laws of perfect gases do not apply. The white dwarfs do not produce energy from nuclear reactions and cool down gradually like a hot metal coming out of a furnace.

b) If the mass of the star is less than $3 M_\odot$, but not too low, the star may condense even more as it collapses. Its density may reach $10^{11} - 10^{14}$ times the density of water. This density is similar to the density inside atomic nuclei. The electrons and protons react and form neutrons. Thus "neutron stars" are formed. These have been studied by *Oppenheimer* and *Volkoff* (1935). The discovery of pulsars (Sect. 1.10) in 1967 by *Hewish* and *S. J. Bell* confirmed their existence.

c) If the star is more massive than $3 M_\odot$, the collapse is complete and leads to a black hole. The whole star enters the horizon and ends up in the singularity at the centre (Fig. 5.6)

The relative sizes of the various stars are shown in Fig. 5.7.

It is likely that massive black holes, with masses of the order of $10^8 M_\odot$, may exist in the centres of galaxies. Some quasar theories, for example, invoke a huge black hole at the centre of the quasar, the result of the concentration of many stars there (*Lynden-Bell*). It is possible that the energy of quasars is produced by stars falling continually into the black hole. It is estimated that the black hole accretes at the rate of one solar mass per year. This theory is considered quite likely today.

5.9 Observations of Black Holes

Many efforts have been made to find black holes in our Galaxy. The most likely candidate is the double star Cygnus X-1 which is an intense x-ray source. The x-rays from this star can only be observed by satellites above the atmosphere since they never reach the surface of the Earth, being absorbed by the Earth's atmosphere. A bright supergiant star has been identified with this source. Its spectrum, however, shows a periodic variation, which means that the star is moving under the attraction of a companion. The companion is invisible, but its mass can be estimated from the perturbations in the orbit of the supergiant. It was found that its mass is between 8 and 18 solar masses. Such a massive star cannot be a white dwarf or a neutron star. Further, if it has not collapsed, it should be very bright. We infer, therefore, that most likely the invisible companion is a black hole.

The x-rays we receive most likely come from particles attracted by the black hole and falling onto an accretion disc around it. The accretion disc is made up of particles orbiting the black hole in circular orbits. As the incoming material falls onto the disc, it loses kinetic energy which is radiated away as x-rays. The observations reveal variations in the x-ray radiation with a period of 0.1 sec. The implication of this is that the region which is emitting the x-rays has dimensions of the order of 0.1 light seconds, that is 30 000 km. Therefore, the star which is in the middle of the accretion disc is smaller than 30 000 km. Stars as small as that are not known to exist, unless they have suffered a collapse. So, these observations support the idea that Cygnus X-1 is a black hole.

So far we have evidence for few other sources of x-rays that are likely to be black holes. One is at the centre of the globular cluster NGC 6624. Two more candidates are the flaring x-ray source AO 620-00 and the x-ray binary LMC X-3 in the Large Magellanic Cloud. Therefore, the question of black holes created by collapsing stars requires further investigation.

We also have evidence for the existence of black holes at the centres of quasars. Indeed, accurate photometric observations of quasars indicate that their nuclei are small, of the order of 10^{10} km in size. Further, certain changes in the quasar luminosities have periods of the order of 1 day, implying that the size of the source is less than one light day ($= 3 \times 10^{10}$ km). Such dimensions are only 50 times the Schwarzschild radius for a body of $10^8 \, M_\odot$. Therefore, it is quite likely that the collapse of the quasar nucleus has formed a large black hole. This question, however, has not been fully resolved yet.

5.10 Rotating Black Holes

The mathematician *Kerr* made one of the greatest discoveries in the field of general relativity in 1963. He found a solution of Einstein's equations which represents rotating black holes. Kerr black holes have certain very important properties, and we shall mention some of them here. Every black hole of this type is surrounded by a spherical event horizon, just like Schwarzschild's black holes. Both horizons have the same properties, that is a particle or a wave crossing it inwards can never come back out. Outside the horizon of the Kerr black hole, there is another ellipsoidal surface which touches the horizon at the poles (Fig. 5.8). It is called the "static limit", because any particle inside it cannot remain still, no matter what other forces act upon it. Instead, it is dragged into motion around the black hole.

Kerr's black hole is characterised by its mass M and angular momentum J. *Newman* generalised this solution by including a charge Q on the black hole. When the angular momentum and the charge become zero, we have Schwarzschild's black hole. In this case, the static limit and the event horizon coincide. When the angular momentum is zero, but the charge is non-zero, the black hole is of the Reissner-Nordström type.

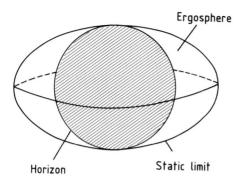

Ergosphere

Fig. 5.8. Kerr black hole

Horizon   Static limit

The region between the static limit and the horizon of a Kerr black hole is called the "ergosphere". It is possible for a particle fallinig inside the ergosphere to break into two parts, one of which will fall into the black hole and the other will come out. The particle that comes out may carry with it more energy than the initial particle. The energy gained comes from the black hole itself. This is the "Penrose mechanism" (*Penrose* 1969) for producing energy. It is the most efficient method for producing energy, being much more efficient than nuclear reactions and even the annihilation of matter with antimatter.

So, some people devised a method by which black holes might be used in the future for energy production. A technologically advanced civilisation could build every city of the future around a black hole (Fig. 5.9). Every day a set of vehicles will carry the rubbish of the city to the ergosphere and throw it into the horizon while the vehicles will gain energy. The vehicles will come out of the ergosphere and return to the city to set in motion its power stations. The amount of energy produced in this way can solve the energy problem for millions, or even billions of years.

Of course, it is not possible to carry on producing energy in this way forever. This is because the energy which may be extracted from a black hole is very high but not infinite. It cannot exceed the rotational energy of the black hole.

Christodoulou has formulated an important theorem concerning the energy of rotating black holes (1970). He distinguished between reversible and irreversible changes of the mass M and the angular momentum J of a Kerr black hole. A reversible change does not alter the surface of the black hole (i.e. the event horizon). The effect of such a change can be reversed so that the quantities M and J may be restored to their original values. On the other hand, an irreversible change increases the surface of the black hole. Similar results hold for the Kerr-Newman black holes which also have charge. In the case of a reversible change, the minimum

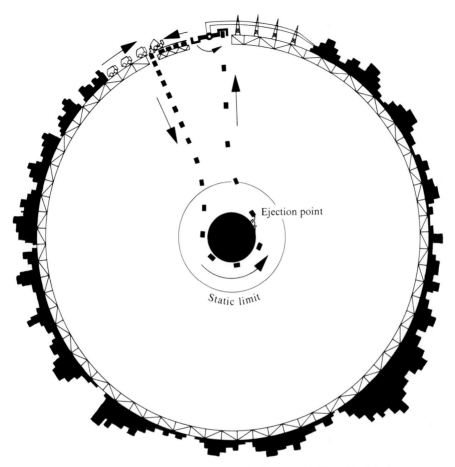

Fig. 5.9. Energy production in an imaginary city built around a Kerr black hole

value of the mass of the black hole is

$$M_\alpha = \frac{c^2}{4G} \sqrt{\frac{A}{\pi}}, \qquad (5.23)$$

where A is the area of the horizon. This formula is derived from (5.16) if we note that $A = 4\pi r_s^2$. M_α is called the "irreducible mass" of a black hole and does not change during a reversible change. However, an irreversible change (caused, for example, by the fall of matter into the black hole) increases M_α. The relation between the total mass M of a black hole, the irreducible mass M_α, its angular momentum J and charge Q is

$$M^2 = \left(M_\alpha + \frac{Q^2}{4M_\alpha}\right)^2 + \frac{J^2}{4M_\alpha^2}. \qquad (5.24)$$

The mass M, therefore, may change if the charge Q and the angular momentum J change, although M_α may not change. For example, if particles with a charge opposite to Q fall into the black hole, the absolute value of Q will be reduced. Similarly the angular momentum J may be reduced if particles with opposite angular momentum fall onto the black hole. By changing J and Q in this way we may change the mass by ΔM to produce an amount of energy equal to $E = \Delta M c^2$. Christodoulou's formula (5.24) allows us to calculate the maximum amount of energy that we can extract from a black hole, which is $(M - M_\alpha) c^2$.

Another important theorem concerning black holes is the "no hair theorem". This means that the only quantities which may characterise a black hole are its mass M, its angular momentum J and its charge Q. The collapse of a star to form a black hole wipes out all other details of its structure, or characteristic quantities. We can never discover any other properties of the star which formed the black hole, for example whether it was made of matter or antimatter, whether it had a magnetic field, or an irregular shape, and so on. None of these characteristics leave any trace outside the black hole, and that is what is meant by "hair". This theorem is very significant, because it severely restricts the final endpoints of stellar evolution and, in general, concentrations of matter. The only final form for such a concentration is a Kerr-Newman black hole (i.e. a rotating charged black hole), and no other options are possible. This is very important for black hole thermodynamics.

5.11 Thermodynamics and Quantum Mechanics of Black Holes

Christodoulou's distinction between reversible and irreversible changes in black holes reminds us of the reversible and irreversible processes in thermodynamics. There is a quantity in thermodynamics which remains constant during reversible processes and increases during irreversible ones without ever decreasing. This is the entropy S.[4] In the case of black holes the quantity which remains constant during reversible processes, while it only increases during irreversible ones, is the area of the horizon A. This led *Bekenstein*, in 1972, to suggest that every black hole has an entropy S which is proportional to the area of its horizon A.

Indeed the entropy of a system containing a black hole increases continuously, due to irreversible processes. The fall of material into a black hole increases its area. Similarly, it has been shown that when two

---

[4] In statistical mechanics the entropy S is related to the probability P that a system will be in a given state, by the relation $S = k \ln P$, where k is the Boltzmann constant. The probability of an isolated system either remains constant or increases in time. It can never decrease.

black holes collide, they form a new black hole, with an area greater than the sum of the areas of the initial black holes.

There is yet another observation that makes more plausible the correspondence between entropy and the area of black holes. The "no hair theorem" shows that the collapse of a star into a black hole results in the loss of a large amount of information about the star. For example, the information we had about the star's structure and the distribution of its matter and radiation is lost, because the only remaining characteristics of the star after its collapse are its mass, its charge (usually zero) and its angular momentum. One might think that the amount of lost information is infinite (the position and velocity of every point in the star). Due to the quantum structure of matter and radiation, however, the amount of our information is finite. That is, it is not possible to have information about regions smaller than the dimensions of the quanta which make up the star. The information we have for the star describes only one of the possible cases of stars which may lead to a similar black hole. If N are all the possible cases, the probability P for a specific star to collapse to form a black hole is $1/N$. This led *Bekenstein* to define the entropy S of a black hole as the logarithm of the probability P (multiplied by Boltzmann's constant) just as in ordinary thermodynamics. This definition establishes a correspondence between S and A. More accurately:

$$S = \frac{k A}{4 \hbar}, \tag{5.25}$$

where k and $\hbar$ are Boltzmann's and Planck's constants respectively.

Bekenstein's definition led, however, to certain basic difficulties. If a black hole has entropy, it has temperature as well. The temperature must be proportional to its surface gravity. The black hole, therefore, acts like a "black body" which absorbs any radiation which falls onto it. A problem arises then from the fact that a black body *radiates* when it is in an environment cooler than itself. For example, the Sun is approximately a black body and is still very bright. So, the black holes should emit radiation which, at first sight, is in contradiction to their definition.

Hawking solved this paradox in 1974, by founding the *quantum mechanics of the black holes*. According to *Hawking*, black holes emit particles and light just like a black body of the same temperature. One way to understand the mechanism for this emission is with the help of Heisenberg's uncertainty principle. Because of this principle, a particle inside the black hole, but close to its horizon, has a certain probability of being outside it, and be observed by a distant observer as if it had been emitted by the black hole. Without the uncertainty principle this phenomenon would have been impossible because the particle ought to be either inside or outside the black hole. This phenomenon, in the case of a system of many particles, gives a percentage of particles inside the black hole, and a percentage outside it.

This is similar to the "tunnel effect" that appears in the radiation from radioactive nuclei. There is a potential barrier around the nucleus which cannot be penetrated by a particle inside it, according to classical mechanics. It is as if somebody is inside a well, drilled on the side of a hill, and tries to jump out. If, when he jumps, he cannot reach the top of the well he can never get out. Quantum mechanics says, however, that there is a small, but non-zero, probability that he may get out without acquiring the energy to jump out of the well. It is *as if* he managed to open a tunnel in the wall of the well, through to the hillside. This is why this phenomenon is called the "tunnel effect".

Something similar happens with the emission of light and particles from a black hole. This is a purely quantum effect and it would not be possible if only classical theory were applicable. A consequence of the Hawking effect is that every black hole continuously diminishes by emitting particles and photons. When its mass approaches zero, the emission process proceeds very rapidly, and finally ends in an enormous explosion when the black hole is destroyed.

Let us now examine the practical significance of the Hawking effect. A black hole, the size of the Sun, has a temperature 10^{-7} K (one tenth of a millionth of a degree above absolute zero). The radiation from such a black hole is entirely insignificant. It is estimated that 10^{66} years would be required for all the mass of such a black hole to be radiated away. Given that ordinary black holes have masses at least equal to $3 M_\odot$, their life must be even longer. Further, the black holes are surrounded by the microwave background radiation (Sect. 3.6) which has temperature 2.7 K, i.e. incomparably higher than the temperatures of the black holes themselves. The black holes, therefore, absorb instead of emitting radiation, just like black bodies. Only in a continuously expanding Universe will the temperature of the microwave background radiation eventually drop below 10^{-7} K, in which case the black holes will start emitting radiation. Even in this case 10^{66} years will have to elapse before the black holes are consumed. The Hawking effect, therefore, is entirely negligible for ordinary black holes formed from collapsing stars.

On the other hand, primordial black holes formed during the Big Bang may be very small, with masses much less than a solar mass. For example, a black hole with mass 10^{15} g (i.e. one trillionth of the mass of the Earth) may be of the size of a proton and have temperature 1.2×10^{11} K. Such a black hole radiates a power of 6000 megawatts, and becomes exhausted after only 10 billion years. Therefore, it is possible even today to observe the final explosions of primordial black holes. These black holes, in their final stages, will emit very energetic γ-rays. Special instruments launched on satellites have failed to detect such γ-rays. This implies that primordial black holes are either very scarce, or they do not exist at all. The subject is very interesting and needs further investigation.

6. Relativistic Cosmology

6.1 Cosmological Solutions of Einstein's Equations

We are going to discuss now the solutions of Einstein's equations which
refer to the whole Universe. The simplest of them refers to a homogeneous
and isotropic Universe.

It is possible to show that for a homogeneous and isotropic Universe
the line element ds^2 can be written as

$$ds^2 = c^2 dt^2 - R^2(t) d\sigma^2, \quad \text{where} \tag{6.1}$$

$$d\sigma^2 = \frac{dr^2 + r^2(d\theta^2 + \sin^2 d\phi^2)}{(1 + \varepsilon r^2/4)^2}, \tag{6.2}$$

where ε can be 0 or ± 1[1]. This line element has been derived by *Robertson*
and *Walker* and holds for every homogeneous and isotropic system, independently of whether Einstein's equations hold or not. $R(t)$ is a function
of time only, called the "scale factor" (or "radius of the Universe") and
the Universe's expansion depends upon it.

The expansion of the Universe does not alter the radial coordinate r of
a galaxy. The true distance of a galaxy, however, is Rr where R increases
with time. So we may say that R is the distance between two galaxies which
are at coordinate distance $r = 1$ away from each other. If $\varepsilon = 0$, the space
is Euclidean, if $\varepsilon = -1$ the space is "hyperbolic", while if $\varepsilon = 1$ it is
"spherical".

Einstein also suggested a more general form for the field equations
(5.11):

$$G_{\mu\nu} + \lambda g_{\mu\nu} = -\kappa T_{\mu\nu}, \tag{6.3}$$

where λ is the "cosmological constant", and $\kappa = 8\pi G/c^4$. Using either

---

[1] If we set $r' = r/(1 + \varepsilon r^2/4)$, $d\sigma^2$ may be written in another commonly used form:

$$d\sigma^2 = \frac{dr'}{1 - \varepsilon r'^2} + r'^2(d\theta^2 + \sin^2 \theta \, d\phi^2).$$

equations (5.11) or (6.3) we find that the function $R(t)$ must satisfy the two conditions (in units such that $c = 1$)

$$\tfrac{4}{3}\pi R^3 \varrho = M = \text{constant} > 0 \quad \text{and} \tag{6.4}$$

$$\left(\frac{dR}{dt}\right)^2 = \frac{2GM}{R} + \frac{\lambda R^2}{3} - \varepsilon. \tag{6.5}$$

These conditions were first derived by *Lemaitre* (1927).

The first condition essentially means that the mass-energy of a sphere of radius R does not change as R increases with the expansion of the Universe, that is, no matter or energy is created out of nothing; instead, the density drops because of the expansion.

The second condition is a simple differential equation which can be easily solved. Every solution of this equation represents a different model of the Universe.

For a static Universe, R does not change with time. It can then be shown that $\varepsilon = 1$, and λ takes a specific value

$$\lambda_c = \frac{64\pi^2}{9\kappa^2 M^2},$$

while the scale factor R is equal to $R_c = 1/\sqrt{\lambda_c}$. This solution was derived by *Einstein* in 1917.

Lemaitre showed that this solution is unstable. Indeed, it is easy to show that if R (or λ) is slightly perturbed away from the value R_c (or λ_c), then dR/dt, as calculated from (6.5), will not remain zero but start increasing (or decreasing) with time at an ever faster rate. That is, the Universe will either start expanding, or contracting, in a continuously accelerated manner.

We are only interested in those solutions of (6.5) which include a stage of expansion of the Universe. There are three types, depending upon the values of ε and λ:

Type I Solution: Continuously Expanding Universe

The function $R(t)$ continuously increases with time (Fig. 6.1). For $t = 0$, $R = 0$, the whole Universe starts from a condition of infinite density. From then on, the Universe expands continuously and will continue to expand forever.

Such solutions are obtained for:

a) $\varepsilon = 0$ and $\lambda > 0$,   b) $\varepsilon = -1$ and $\lambda > 0$,   c) $\varepsilon = 1$ and $\lambda > \lambda_c$.

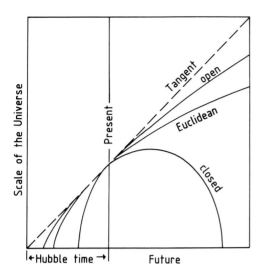

Fig. 6.1. Cosmological models of two ever expanding and a pulsating Universe

Type II Solution: Pulsating Universe

The function $R(t)$ increases initially from zero up to a maximum value $R_{\max}$ and afterwards decreases to zero again (Fig. 6.1). The Universe initially expands, and then contracts, i.e. it is pulsating.

Such solutions are obtained for:

    a) $\varepsilon = 0$ and $\lambda < 0$,   b) $\varepsilon = -1$ and $\lambda < 0$,   c) $\varepsilon = 1$ and $\lambda < \lambda_c$.

Type III Solution: A Universe That Does not Go Through a Singularity at $R = 0$

When $\varepsilon = 1$ and $0 < \lambda < \lambda_c$, there is a solution according to which the Universe initially contracts without ever reaching the singularity $R = 0$, and afterwards it expands. This solution is not particularly interesting because it does not include a superdense state, contrary to the strong evidence we have for such a stage in the past history of the Universe (Sect. 8.1).

Besides the above solutions, we also have the limiting case where $\varepsilon = 1$ and $\lambda = \lambda_c$ (Einstein's static solution). Further, there are two more solutions: (a) The *Lemaitre-Eddington* Universe which starts like Einstein's solution and carries on with a continuous expansion, and (b) a solution which starts with $R = 0$ and reaches asymptotically Einstein's solution in infinite time.

When R increases unrestrictedly, the density of the Universe tends to zero. In the case of an entirely empty Euclidean Universe ($\varrho = \varepsilon = 0$), we have an exponential expansion

$$\frac{dR}{dt} = \sqrt{\frac{\lambda}{3}} R, \quad \text{therefore,}$$

$$R = R_0 \exp\left(\sqrt{\frac{\lambda}{3}} t\right). \tag{6.6}$$

This solution was derived in 1917 by *de Sitter* and it was the first expanding model of the Universe. On the other hand, the first model of a non-empty expanding Universe was derived in 1922 by *Friedmann*, who used the values $\varepsilon = 1$ and $\lambda = 0$.

We note that *Einstein* added the cosmological constant λ to his equations because he wanted to find a static solution of (6.5) and there is no static solution for $\lambda = 0$. Later on, it was shown that the Universe expands and the cosmological constant was not necessary any more. So, from then on most cosmologists, *Einstein* included, used the initial form of the field equations (5.11), that is, they accepted $\lambda = 0$. The models derived in this way are called Friedmann models, or Lemaitre models with $\lambda = 0$.

For $\lambda = 0$ we have the following possibilities:

1. The Universe is continuously expanding if $\varepsilon = 0$ or -1.
2. The Universe is pulsating if $\varepsilon = 1$.

We see, therefore, that for $\lambda = 0$ the geometry of the Universe (i.e. whether the Universe is Euclidean, hyperbolic or spherical) is directly related to whether the Universe is expanding continuously or pulsating.

The scale factor R is plotted against the time t in Fig. 6.1. The spherical Universe is pulsating, while the Euclidean and the hyperbolic Universes are continuously expanding. We take the moment when $R = 0$ as the origin of time.

All three curves have the same tangent (dashed line) at the present time, because the observations give the current rate of the expansion (dR/dt). If dR/dt did not vary with time, the dashed line would represent the expansion of the Universe. Then the time which would have elapsed since $R = 0$ would be $t = 1/H$, where H is Hubble's constant (Sect. 3.7) which enters into Hubble's law, written in the form (6.19). This time is the so called "Hubble time". The age of the Universe must be less than the Hubble time, but of the same order of magnitude.

6.2 The Geometry of the Universe

As we have seen, a homogeneous and isotropic Universe can have one of the following two forms:

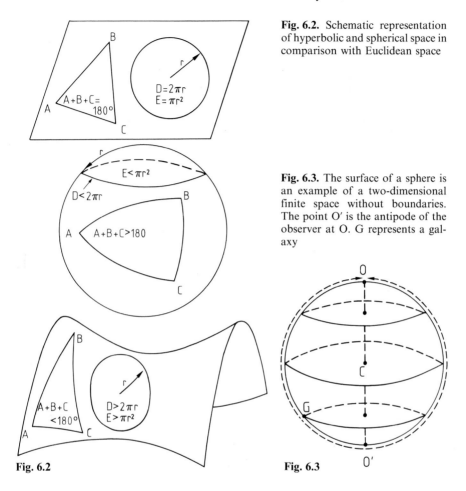

Fig. 6.2. Schematic representation of hyperbolic and spherical space in comparison with Euclidean space

Fig. 6.3. The surface of a sphere is an example of a two-dimensional finite space without boundaries. The point O′ is the antipode of the observer at O. G represents a galaxy

Fig. 6.2

Fig. 6.3

a) it may be hyperbolic ($\varepsilon = -1$), and continuously expanding, or
b) it may be spherical ($\varepsilon = 1$), and pulsating.

The case of the Euclidean Universe ($\varepsilon = 0$), which is continuously expanding, is considered as a special case of the hyperbolic Universe. The Universe is infinite in the first case, and finite without boundaries in the second.

The space in the Universe is three-dimensional and it may be hyperbolic or spherical. Since it is not possible to give a picture of a hyperbolic or spherical space of three dimensions, we shall use the corresponding examples in two dimensions. These are simple surfaces in the usual space, and we shall compare them with an ordinary plane, which could represent a Euclidean Universe (Fig. 6.2).

The sum of the three angles of a triangle on a plane is 180°. Also, the circumference of a circle of radius r is $2\pi r$ and the area of the circle is πr^2.

In a hyperbolic space, the sum of the angles of a triangle is less than 180°, the circumference of a circle is more than $2\pi r$ and its area more than πr^2. On a spherical surface, on the other hand, the sum of the angles of a triangle is more than 180° (e.g. in the case of a three-rectangular triangle, the sum is 270°), the circumference is less than $2\pi r$ and the area less than πr^2. The radius r in these cases is an arc of a geodesic measured on the curved surface. We note that if we start from a certain point O on the surface of a sphere of radius R, and draw circles of increasing radius r, the circumference keeps increasing until the circle becomes a great circle of the sphere, and then starts *decreasing* until r becomes equal to πR, when the circumference is zero (point O' in Fig. 6.3). The point O' is the antipode of O. The maximum value attained by the circumference is $\Pi = 2\pi R$ for $r = \pi R/2$. We notice that $\Pi = 4r$ instead of $2\pi r$, as in Euclidean space. If we repeat the exercise and measure the area of the circle instead of its circumference, we shall see that it increases with r. When $r = \pi R/2$, the circle covers half the surface of the sphere, therefore its area is $2\pi R^2$, i.e. $8r^2/\pi$, instead of πr^2 for a Euclidean space. When $r = \pi R$, i.e. when we have reached the antipode O', the circle covers the whole sphere and therefore its area is $4\pi R^2$. We note that the area of the spherical surface is finite, although the surface does not have any boundaries. So we have a space which is finite without boundaries.

The situation is similar in a non-Euclidean space of three dimensions (Sect. 6.3). More specifically, in a spherical space, the area of a sphere of radius r is less than $4\pi r^2$ and its volume less than $4\pi r^3/3$.

In this case, the *area* of the sphere increases until it reaches a maximum value and then decreases until it reaches the value zero at the antipode. Further, the volume of a sphere of radius πR (the distance to the antipode) is equal to $2\pi^2 R^3$. This volume is the volume of the whole spherical model of the Universe.

If we proceed further than the antipode on the spherical surface (Fig. 6.3), the area does not increase because we simply cover regions we had covered before. The same happens in the case of a three dimensional spherical space. If we try to calculate the volume of a sphere with radius greater than πR, this volume will include regions which had been included earlier. This can be better explained in terms of galaxy counts. If we can see galaxies at distances greater than πR, we shall simply see galaxies which have already been counted.

The volume of the Universe, therefore, is equal to $2\pi^2 R^3$, but this volume does not have boundaries. There is no surface which envelopes it which would correspond to the boundary of the Universe. The "spherical Universe" is not like the *volume* of a sphere, limited by its surface, in spite of the impression to the contrary created by the use of the word "spherical". The spherical Universe is only similar to the *surface* of a sphere, with the difference that the Universe is three-dimensional, while the surface of an ordinary sphere is two-dimensional.

Many people ask what is outside this finite Universe without boundaries. This question is meaningless, because there is no space outside the Universe to discuss what is there.

A property of a finite space without boundaries is that we may traverse it and come back to where we started from, by continually moving in the same direction. We may understand this by going back to our analogy, the surface of a sphere. The "straight lines" (geodesics) on this surface are arcs of great circles. We may imagine that on the surface of a sphere live two-dimensional creatures which do not have the concept of the third dimension. These creatures move along arcs of great circles when they say that they move "straight ahead". These arcs are the shortest routes between two points. A traveller can, therefore, move through the antipode and return back to his starting point without ever having to turn left or right. In a similar way, if our Universe is spherical, we are able to come back to the point we started from by simply moving continuously in the same direction, without deviating to the left or right. We cannot conceive *how* this is realised because we are *not outside* the three-dimensional spherical space. Our difficulties in this respect are the same as those of the two-dimensional creatures on the surface of the sphere. They cannot understand *how* they returned to their starting point, because they cannot see the sphere in the extra (third) dimension, as we can.

Another property of finite spaces without boundaries is that we are able to see the same object in two opposite directions in the sky. Indeed, the geodesic which connects the observer at O to a galaxy G continues round the whole Universe and returns to the observer (Fig. 6.3). So, if T light years is the periphery of the Universe, and τ light years is the shortest distance to the galaxy, the same galaxy can be seen either at a distance of τ, or at $T - \tau$ light years in the opposite direction.

An observer, therefore, must be able to see himself at a distance of T light years if he looks in any direction. More accurately, he must be able to see how the Earth was T years ago. Indeed, the light which set off from the Earth T years ago, is being focussed back towards the Earth again. The time T, however, is very large, at least of the order of 20 billion years. Therefore, if we could really see objects as far as that, we wouldn't be able to see the Earth or the Sun, because neither of them existed at that time; only the matter out of which they are made existed. What is more, all this could only happen if T is less than the age of the Universe, otherwise the light would not have had time to go round the Universe once, to cause these strange effects.

The basic question we have not addressed, however, is whether the space is in reality infinite or finite without boundaries. In order to answer this question, we must again turn to the observations (Sect. 6.6).

6.3 Spherical Space of Three Dimensions

The line element of a three-dimensional spherical space is given by

$$d\sigma^2 = \frac{dr^2 + r^2(d\theta^2 + \sin^2\theta\, d\phi^2)}{(1 + r^2/4)^2}.$$ (6.7)

Here r expresses the "co-ordinate radial distance" while the usual radial distance is Rr. Similarly, the real distance between two neighbouring points is $Rd\sigma$ and not $d\sigma$. We may now introduce a new variable x, instead of r, such that

$$r = 2\tan\frac{x}{2}.$$ (6.8)

Then we find that

$$d\sigma^2 = dx^2 + \sin^2 x(d\theta^2 + \sin^2\theta\, d\phi^2).$$ (6.9)

The area of an elementary region between the angles θ and $\theta + d\theta$ and ϕ and $\phi + d\phi$ is

$$dE = R\,\sin x\, d\theta\, R\,\sin x\,\sin\theta\, d\phi = R^2\sin^2 x\,\sin\theta\, d\theta\, d\phi,$$ (6.10)

while the elementary volume with base area dE and height $R\,dx$ is

$$d\Omega = R^3\sin^2 x\, dx\,\sin\theta\, d\theta\, d\phi.$$ (6.11)

If we integrate the elementary area from $\phi = 0$ to $\phi = 2\pi$ and from $\theta = 0$ to $\theta = \pi$, we find the area of a sphere of radius x

$$E = 4\pi R^2\sin^2 x,$$ (6.12)

while, if we integrate $d\Omega$ over the same limits of ϕ and θ and x from 0 to x, we find

$$\Omega = 2\pi R^3\left(x - \frac{\sin 2x}{2}\right).$$ (6.13)

The area E is less than the area of a sphere in Euclidean space, $4\pi R^2 x^2$. We also notice that E reaches a maximum value for $x = \pi/2$ and zero for $x = \pi$.

Further, the volume (6.13) is less than the volume of a sphere in a Euclidean space, which is $4\pi R^3 x^3/3$.

When we reach the antipode, for $x = \pi$, we have the "total volume of the spherical space", which is:

$$\Omega = 2\pi^2 R^3. \tag{6.14}$$

6.4 The Theory of the Expanding Universe

The expansion of the Universe is the most important cosmological event. The whole Universe is expanding. The distant galaxies move away from each other with relative velocities which increase as their separations increase (Fig. 6.4). Thus, the expansion of the Universe is like an enormous explosion. In the case of the explosion of a bomb, the various fragments move away from the centre very rapidly, and the higher their velocities are, the further away they will reach. So the relative velocity of the fragments increases with their separation. Something similar happens in the Universe. The only difference is that in the Universe there is no particular point which could be considered as the "centre of the explosion". This is because the whole of the Universe participates in the explosion. Further, the fragments in the explosion of a bomb fill up more and more space, while in the Universe there is no space which is not already filled with galaxies. We may say that the galaxies are almost fixed in space, and that the space itself is expanding. In mathematical terms, this is expressed by the fact that the co-ordinates of a galaxy (r, θ, ϕ) do not

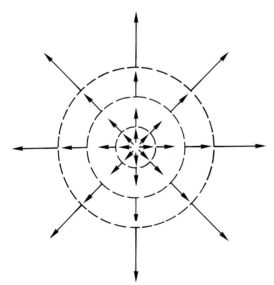

Fig. 6.4. Expansion of the Universe

change, but the scale factor R changes. Something similar happens with a balloon which represents the Earth, and is continuously inflated. The various points on the surface of the balloon have constant co-ordinates (longitude and latitude) but their distances continuously increase.

The expansion of the Universe is detected by observing the spectra of distant galaxies (Sect. 3.7). We find that the spectral lines of these galaxies are redshifted. If λ is the wavelength of a spectral line, the relative shift of the line is $z = \Delta\lambda/\lambda$ and is given by Doppler's formula

$$z = \frac{v}{c}, \tag{6.15}$$

where v is the velocity of the source relative to the observer. When v approaches the speed of light c, we use the more accurate relativistic formula[2]

$$z = \sqrt{\frac{1 + v/c}{1 - v/c}} - 1. \tag{6.16}$$

Hubble was the one who studied most systematically the radial velocities of galaxies and found empirically the fundamental law of the cosmic expansion:

$$cz = Hr, \tag{6.17}$$

where r is the "luminosity distance" of a galaxy and H is a constant called the "Hubble constant".

Observations indeed show a relation between bolometric magnitude m of a galaxy and its redshift. In Fig. 3.7, the observational results are shown as points and they are plotted with the theoretical curves A, B, C, D which represent the various models of the expanding Universe. It seems that the model which fits the observations best is either model B or model C.

When the distances between galaxies are relatively small, according to formula (6.15), we have

$$v = Hr \tag{6.18}$$

that is, the velocity of the expansion is proportional to the distance. If we use the scale factor R, we may write

$$\frac{dR}{dt} = HR. \tag{6.19}$$

---

[2] For example, for $z = 1$, $v = 0.6c$ rather than $v = c$ as the ordinary Doppler formula would have given. If $z = 2$, $v = 0.8c$ and if $z = 3.5$, $v = 0.91c = 273\,000$ km/sec.

When the distance r is large, we use the more accurate formula (6.30). In any case, the recession velocity always increases with the distance.

The redshift z is easily measured from the spectra of the galaxies. It is more difficult, however, to measure the distance r, and thus to calculate H. Initially, *Hubble* estimated that $H = 530$ km/sec/Mpc. Today, after several revisions, we accept that H is between 50 and 100 km/sec/Mpc (Sect. 3.7).

6.5 Theoretical Interpretation of Hubble's Law

To explain the Hubble law (6.17), we start by making the observation that the light which comes from a galaxy follows a geodesic along which $ds = 0$ and $d\theta = d\phi = 0$. By using formulae (6.1) and (6.2) we can then obtain

$$0 = c^2 dt^2 - \frac{R^2(t)\, dr^2}{(1 + \varepsilon r^2/4)^2} \quad \text{or} \tag{6.20}$$

$$\frac{dt}{R(t)} = -\frac{dr}{c(1 + \varepsilon r^2/4)}. \tag{6.21}$$

We have kept the minus $(-)$ sign because the light comes towards us, i.e. propagates to smaller r.

As we mentioned earlier, the co-ordinate distance r of a galaxy does not change with time. Therefore, the sum of the quantities $dt/R(t)$ is constant and independent of the expansion of the Universe, from the moment that the light sets off from the galaxy (at time t), until the moment it reaches the Earth (at time t_0).

If a second light ray sets off from the galaxy after time $\Delta t = \lambda/c$ (where λ is the wavelength of the emitted light) it arrives at the Earth at time $\Delta t_0 = \lambda_0/c$ later than t_0. The corresponding sum of the quantities $dt/R(t)$ is the same as before. The two sums, however, are different by $\Delta t/R(t)$ at the beginning of the route and $\Delta t_0/R(t_0)$ at the end. This means that these two quantities are equal. That is

$$\frac{\Delta t}{R(t)} = \frac{\Delta t_0}{R(t_0)}. \tag{6.22}$$

Therefore, the wavelengths λ, λ_0 and the corresponding frequencies v, v_0 are related by

$$\frac{\lambda}{\lambda_0} = \frac{v_0}{v} = \frac{R(t)}{R(t_0)}. \tag{6.23}$$

The relative redshift is equal to

$$z = \frac{\Delta\lambda}{\lambda} = \frac{\lambda_0 - \lambda}{\lambda} = \frac{R(t_0)}{R(t)} - 1. \tag{6.24}$$

Assuming that $(t_0 - t)$ is small, we may expand $R(t)$ to find

$$R(t) = R_0 - \dot{R}_0(t_0 - t) + \cdots \tag{6.25}$$

where we have put R_0 for $R(t_0)$ and $\dot{R}_0$ means the derivative of R with respect to time, calculated at $t = t_0$. From the last two equations we find

$$z = \frac{\dot{R}_0}{R_0}(t_0 - t) + \cdots. \tag{6.26}$$

We also have, approximately, that

$$r = c(t_0 - t) + \cdots. \tag{6.27}$$

Therefore,

$$cz = Hr + \cdots, \tag{6.28}$$

where we have put

$$H = \frac{\dot{R}_0}{R_0}. \tag{6.29}$$

Equation (6.28) is Hubble's law in its simplest form. If we keep up to second order terms in the various expansions we obtain

$$cz = Hr + \frac{H^2 r^2}{2c}(q - 1), \tag{6.30}$$

where the quantity

$$q = -\frac{R_0 \ddot{R}_0}{\dot{R}_0^2} \tag{6.31}$$

is the so-called "deceleration parameter", since it is positive when the second derivative of R is negative ($\ddot{R} < 0$), i.e. when the expansion is decelerated. Equation (6.30) expresses Hubble's generalised law.

If the cosmological constant λ is zero, (6.4) and (6.5) give

$$q = \frac{4\pi G\varrho}{3H^2} \quad \text{and} \tag{6.32}$$

$$\varepsilon = \frac{8\pi G \varrho R^2}{3q}\left(q - \frac{1}{2}\right). \tag{6.33}$$

The Universe, therefore, is spherical if $q > 1/2$ (i.e. $\varepsilon > 0$) and hyperbolic if $q < 1/2$ (i.e. $\varepsilon < 0$). However, q is related to the density of the Universe ϱ through (6.32). Therefore, if ϱ is small, q is small, and the Universe is hyperbolic and expands for ever. If ϱ is above a certain critical value, the Universe is spherical and pulsating. The critical value of ϱ corresponds to $q = 1/2$ and is

$$\varrho_c = \frac{3H^2}{8\pi G}. \tag{6.34}$$

If we put $H = 50$ km/sec/Mpc in the above formula, we get

$$\varrho_c = 5 \times 10^{-30} \text{ g/cm}^3. \tag{6.35}$$

6.6 Is the Universe Finite or Not?

In principle, we can answer this question by calculating the "deceleration parameter" q from the redshift of the galaxies of known distance, and by using Eq. (6.30). If $q \leq 1/2$ the Universe is infinite, otherwise it is finite.

In practice it is very difficult to measure the distances of the far away galaxies, so the calculation of q from (6.30) is very inaccurate. The observational data are not accurate enough to show any meaningful deviation from the straight line defined by the initial Hubble law (6.17). We may only say that q is less than 1.5 and greater than zero. So, by looking at Fig. 3.7 we can exclude the value $q = -1$ which corresponds to the theory of continuous creation (Sect. 7.3). We cannot be sure, however, whether q is less than or greater than 1/2. Thus this method cannot be used in practise to answer the question of whether the Universe is finite or not.

For this reason we try to find q indirectly by measuring the density ϱ and using Eq. (6.32). That is, we try to see if the density of the Universe is above or below the critical density ϱ_c which is necessary for the Universe to "close".

We may say that if $\varrho > \varrho_c$, the gravitational attraction of the matter in the Universe is enough to strongly decelerate its expansion, and eventually to stop it and cause a contraction, leading to a pulsating Universe. On the other hand, if $\varrho < \varrho_c$ the gravitational attraction is inadequate to stop the expansion which will continue forever. This phenomenon is similar to the ejection of a particle from a celestial body. For example, if a projectile is ejected from the Earth with velocity 10 km/sec, it will not escape to infinity but will fall back to Earth. On the other hand, if the same projectile is ejected with the same velocity from the Moon, it will escape to infinity[3]. The gravitational attraction of the Earth is enough to reverse the direction of motion of the projectile, while the attraction of the Moon is inadequate.

Indeed we may derive Eq. (6.34) for the critical density, from Eq. (5.17) for the escape velocity from a body of mass M. Equation (5.17) can be written as

$$v^2 = \frac{2GM}{r}. \tag{6.36}$$

If v is given by Hubble's law $v = Hr$, and the mass M by $M = 4\pi r^3 \varrho_c/3$, we obtain

$$\varrho_c = \frac{3H^2}{8\pi G}, \tag{6.37}$$

which is exactly (6.34).

If the density is greater than ϱ_c, the expansion rate is less than the escape velocity. Then the expansion cannot continue forever and the Universe will pulsate. That is, the critical density ϱ_c is the maximum density which allows the expansion to carry on forever.

As we saw earlier, however, for zero cosmological constant, a pulsating Universe is finite without boundaries while a continuously expanding Universe is infinite. Therefore, to answer the question of whether the Universe is finite or infinite we must determine its density ϱ.

To do so, we first calculate an average galactic mass and divide by the volume inside which there is, on average, one galaxy. Thus, we find that the density of matter in galaxies is only four hundredths of the critical density. When the observational errors are taken into account, one may increase this figure by a factor of three but no more. That is, the maximum possible density of matter in galaxies is only $0.12 \varrho_c$, i.e. eight times less than the critical density.

Is it possible, however, for much more matter to exist between galaxies, in a very diffuse form so that it cannot be observed?

Various propositions have been made for such diffuse matter. This matter could be (a) diffuse intergalactic matter (gas or dust); (b) isolated stars, including dead stars, white dwarfs, neutron stars and black holes; (c) photons, gravitons (corresponding to gravitational waves) or neutrinos; (d) superheavy magnetic monopoles predicted by the "Grand Unified Theories" as well as other exotic particles.

---

[3] We neglect here the attraction of the Sun and the other planets.

6.6.1 Intergalactic Gas and Dust

The density of the intergalactic gas (Sect. 3.1) can be found from observations of the 21 cm-line of neutral hydrogen. The fact that no appreciable radiation has been observed at this wavelength puts an upper limit on the amount of intergalactic gas. This upper limit is compatible with a cosmic density of 0.3 ϱ_c although this is likely to be an overestimate.

The quantity of intergalactic dust is most probably minimal judging from the amount of interstellar dust, which is only 1% of the interstellar gas. If the density of the intergalactic dust were high, it would appreciably absorb the light from distant galaxies, so that their apparent number would decline with distance, and the Universe would appear to be inhomogeneous.

6.6.2 Intergalactic Stars

Very few stars have escaped from the galaxies since the relaxation time is very long for these systems, of the order of 10^{12} years. In 10^{10} years, therefore, only about 1% of the stars are expected to have escaped into the intergalactic medium. Tidal forces between galaxies may have also caused stars to escape from the galaxies, but this is not an appreciable contribution. The only possibility that we are left with, is that intergalactic stars or black holes, which were formed during the first stages of the evolution of the Universe, have not been segregated into galaxies and clusters of galaxies. Indeed, some people have suggested that the space between galaxies is full of mini black holes which have been created in the early Universe and spread throughout space.

The existence of galaxies and clusters of galaxies, however, implies that any form of matter is subject to the law of gravitational attraction and tends to form agglomerations. The various theories of galaxy formation out of small primordial perturbations to the cosmic density, depend upon this property of matter. The tendency to form agglomerations is characteristic of *all* forms of matter, provided that the velocities are not too high. That is, it is not possible for one form of matter (e.g. gas or black holes) to avoid segregation into clusters of galaxies. That is why, we accept that the agglomerations of matter we see in the Universe represent the invisible matter too, such as diffuse matter, dead stars and primordial black holes. As we discussed in Sect. 2.11, the masses of clusters of galaxies can be estimated from observations. For the reasons just stated, we expect that these measurements give us a good estimate of the total mass in the Universe.

6.6.3 Photons and Gravitons

The arguments concerning the segregation of matter into clusters do not hold for photons, gravitons and neutrinos (if their rest mass is zero). These particles move with the speed of light and they cannot form clusters. Although light is affected by gravitational fields, the effect is very small. Therefore, the light does not concentrate in galaxies and clusters of galaxies.

We can estimate the average density of photons in the Universe. Most of the diffuse photons form the microwave background radiation. Their total contribution to the energy, and hence the matter, of the Universe is insignificant. It is estimated that the microwave background radiation corresponds to a density of 3×10^{-34} g/cm^3, i.e. one ten thousandth of the critical density. Similarly, the contribution of gravitons is estimated to be very low.

6.6.4 Neutrinos

Until recently, it was thought that neutrinos have zero rest mass and move with the speed of light. In recent years, however, it has been suggested that neutrinos have a non-zero rest mass and therefore move slower than the speed of light. Various experiments conducted after 1979 have shown evidence that the mass of the neutrino could be $20-30$ eV/c^2.[4] If this is right all the neutrinos in the Universe could have mass well above the total mass of all other forms of matter and energy.

The existence of these "heavy neutrinos" may explain the "missing mass" from the clusters of galaxies, as we saw in Sect. 2.11. Indeed, if enough neutrinos have been concentrated in the clusters of galaxies it is possible to provide the necessary gravitational attraction which will hold the member galaxies together, in spite of their high velocities.

A recent study by *Schramm* and *Steigman* (1981) discusses the matter in detail. The more massive the neutrinos are, the lower their velocities and the easier it is to condense close to the various matter concentrations in the Universe. For example, if the mass of the neutrinos is more than 4 eV/c^2, their velocities are approximately 10 km/sec or less, and they may form most of the mass in clusters of galaxies. If their mass is more than 12 eV/c^2 they may cluster on even smaller scales, i.e. binary galaxies, and if they have masses more than 22 eV/c^2 they may cluster on galactic scales and form the invisible haloes around the galaxies.

Although it is known that there are dark haloes around various galaxies, in most cases their masses can be accounted for by known forms of matter (i.e. faint stars and gas which extend to large distances from the galactic centre). So, it is not likely that the neutrino mass is as high as 22 eV/c^2. On the other hand, if the neutrino mass is less than 4 eV/c^2,

---

[4] The electronvolt (eV) is a unit of energy. The corresponding mass can be found from the formula $m = E/c^2$. It turns out that 1 eV/$c^2 = 1.8 \times 10^{-33}$ g.

neutrinos cannot account for the missing mass in clusters of galaxies. So, *Schramm* and *Steigman* concluded that the neutrino mass must be between 4 and 20 eV/c^2.

This mass is high enough for the majority of matter in the Univese to be in the form of neutrinos. If neutrinos fill up the whole of space the average density of the Universe approaches the critical density ϱ_c and it may even be higher than ϱ_c, so that the Universe will be closed, i.e. finite without boundaries.

The question of the neutrino mass has attracted considerable interest from a large number of researchers, both theoretical and experimental. Although the first experiments indicated a mass of $20-30$ eV/c^2, more recent experiments indicate a mass less than 1 eV/c^2. If neutrinos have non-zero mass, it is possible that there are "neutrino oscillations" between the various types of neutrinos. For example, electron neutrinos may change into muon neutrinos or τ neutrinos (Sect. 8.5), and vice versa. If this is so, it is possible to explain the paradox of the low number of neutrinos we observe coming from the Sun contrary to the theoretical predictions. Systematic experiments conducted at 1500 m below the surface of the Earth have shown that the neutrinos we receive from the Sun are only a third of the theoretically predicted number. Our experiments, however, only detect electron neutrinos produced by the nuclear reactions in the Sun. If neutrinos can change type by such neutrino oscillations, many of them can escape detection. If there are three types of neutrinos, as we accept today (electron, muon and τ neutrinos), and if the neutrino flux from the Sun is in these three forms, then it is clear that we detect only one third of them. This problem, however, requires further research.

6.6.5 Magnetic Monopoles and Other Exotic Particles

Contrary to the usual magnets, which have two inseparable poles, the magnetic monopoles are particles which have only one magnetic pole. Their existence is predicted by the "Grand Unified Theories" (Sect. 8.5). Their mass could be very large, of the order of 10^{15} GeV/c^2 (that is 2×10^{-9} g) and they may contribute significantly to the mass of the Universe. However, such particles have not been observed up to now. Many efforts to find magnetic monopoles in cosmic rays gave negative results. A single claim of such an observation in 1975 is generally doubted.

Other exotic particles predicted by the most recent theories are called photinos, gravitinos, Higgs' bosons, axions, etc. For example, the photino is introduced by the modern supersymmetry theories (Sect. 8.7) as a massive partner of the photon (which has zero rest mass). Such particles, if they exist, may increase the mean density of the Universe. However, it is too early to provide good limits for the masses and the number density of them.

6.6.6 Limits Imposed by Cosmic Nucleosynthesis

We have some indirect evidence which suggests that the density of matter in the Universe is less than the critical density ϱ_c. It comes from the formation of the elements, and in particular deuterium. Deuterium and helium were principally formed during the first four minutes of the Universe, when its temperature and density were very high (Sect. 6.7, Fig. 6.5).

The amount of deuterium produced depends upon the density of matter. The higher the density is, the more helium $_2He^4$ is produced, and the less deuterium is left over. For example, if the density were ten times higher, most of the deuterium would have been converted into helium. Therefore, we can estimate the density of the Universe at the time of cosmic nucleosynthesis (e.g. when its temperature was 10^9 K) by observing the amount of deuterium that exists today. We can then estimate the present density of the Universe, taking into account the change in the scale factor.

Deuterium observations were made by the Copernicus satellite in 1973. It was deduced that the amount of deuterium is 2×10^{-5} times the amount of hydrogen. From this figure one finds that the present density of the Universe is only one tenth of the critical density. This is the strongest argument in support of an infinite Universe which will expand forever.

The subject, however, is not closed yet. Some people doubt the accuracy of the deuterium measurement. So, the question of whether the Universe is infinite or finite is still open. In what follows we shall discuss both possibilities, i.e. that the Universe is infinite or finite without boundaries.

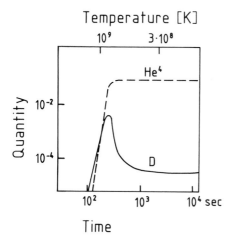

Fig. 6.5. Formation of deuterium (D) and helium ($_2He^4$) during the first stages of the cosmic expansion

6.7 The Big Bang Theory

The most important current theory for the origin of the Universe is the Big Bang theory. We accept that the Universe started with a huge explosion from a superdense and superhot stage. Theoretically, the Universe started from a mathematical singularity with infinite density. This comes out from solutions of type I or II of Einstein's equations (Sect. 6.1). All of them start from a point where the scale factor R is zero at time $t = 0$. Further, the derivative of R is infinite at this time. That is, the initial explosion happened with infinite velocity. Of course, it is impossible for us to picture the initial moment of the "creation" of the Universe. Later on we shall discuss the various current attempts to understand the conditions which prevailed in the first moments of the Universe.

The first scientists to formulate the Big Bang theory were *Lemaitre* in 1927 and *Gamow* in 1948. *Lemaitre* based his conclusions not only on Einstein's equations, but on the following argument as well. The entropy of the Universe increases with time. Therefore, there must have existed a condition of minimum entropy where matter had the maximum possible organisation. So, *Lemaitre* proposed the idea of the "primordial atom" which contained all the matter of the Universe. The explosion of this "atom" produced all the stars and galaxies which make up the Universe today. This argument about the entropy of the Universe is very important. Indeed, the increase of the entropy is one of the fundamental characteristics of the Universe, particularly stressed in recent years (Sect. 9.1).

Gamow (1948) proceeded to investigate the characteristics of the superdense condition of the first moments of the Universe. He concluded that the temperature must have been enormous at this stage. Under these conditions, the protons and neutrons must have formed the various chemical elements. This is the "α-β-γ theory" from the initials (in Russian) of the people who developed it (*Alpher, Bethe* and *Gamow*). It satisfactorily explains the formation of deuterium and helium (Fig. 6.5).

This theory was much more ambitious. It also tried to explain the formation of all the elements up to uranium, by continuously adding neutrons to more and more complex nuclei. It soon became clear, however, that since there is no stable nucleus with mass number 5, the formation of the elements must have stopped at helium. Thus, cosmic nucleosynthesis was necessarily restricted to light elements, up to helium $_2\text{He}^4$. After this failure, the theory was put aside altogether. The next theory about the formation of the chemical elements was the B^2FH theory, named so after *G. Burbidge, M. Burbidge, Fowler* (Nobel prize, 1983) and *Hoyle*. According to this theory, all the elements have been formed in stellar interiors. In particular the heavier elements were formed during supernova explosions.

Salpeter had already shown, in 1952, that three atoms of helium $_2\text{He}^4$ may combine to produce carbon, and other heavier elements by addition of further helium atoms, in the interiors of stars rich in helium, where temperatures may reach 10^8 K. More heavy elements may be formed during the final stages of stellar evolution, when a supernova explosion may occur.

The B^2FH theory was very successful. However, as *Hoyle* and *Tayler* realised in 1964, it could not account for the amount of helium observed in stars, which constitutes 25% of their mass. According to the theory of nuclear reactions in stellar interiors, only 1–4% of the total amount of matter can be helium, which is 6–25 times less than the observed amount. This result made *Hoyle* go back to the formation of elements in the Big Bang theory. Thus, in 1967, *Wagoner, Fowler* and *Hoyle* calculated once again the amount of helium which may be formed in the early Universe, and came to the same conclusions as *Gamow* and his collaborators.

The situation is similar for deuterium, which is observed to have a cosmic abundance of 2×10^{-5}. This amount is much more than the amount expected to form in stars, because deuterium is not expected to survive for very long in stars. The deuterium we observe today, therefore, must have been formed in the early Universe.

So, we conclude that there are two ways by which the elements of matter were formed. According to the first way, the cosmological one, only the light elements (mainly deuterium and helium) were formed during the first four minutes after the Big Bang. The elements which are heavier than helium were formed later on in the interiors of stars. This secondary process for generating elements started as soon as the first stars were formed, and continues today.

Another indication in support of the Big Bang theory is the estimated age of the Universe (Sect. 6.9). Independent estimates of the age of the Universe, based on the expansion of the Universe, the age of the oldest stars in the galaxies, or the ages of the radioactive elements, give numbers of the same order of magnitude. All three methods agree that the age of the Universe is between 10 and 20 billion years. If the Universe did not have a beginning, there would not be an a priori reason for such a good agreement between these three different calculations.

Finally, an important piece of evidence for the Big Bang is the microwave background radiation. This radiation (Sect. 3.6) comes to us from all directions with the same intensity (isotropically), and corresponds to the radiation from a black body at a temperature of approximately 3 K. The radiation is uniformly distributed in space, and does not appear to be clumpy, unlike the distribution of matter. The only credible explanation for this radiation is that it consists of the photons which filled the Universe during the "radiation era", early in the cosmic history. No other plausible explanation has so far been suggested.

Recapitulating, we note that the basic evidence in support of the Big Bang theory is:

a) The solutions of Einstein's equations.
b) The observed helium and deuterium abundance.
c) The agreement between the various independent estimates of the age of the Universe.
d) The microwave background radiation.

In the next section (Sect. 6.8) we shall discuss some extra evidence based on the distribution of radio galaxies.

Of course, none of the above mentioned arguments in support of the Big Bang theory became immediately accepted. In particular, the solutions of Einstein's equations referred to homogeneous and isotropic models of the Universe. So the question arises, what happens if the Universe is not entirely homogeneous and isotropic after all. Is it possible, in that case, to avoid the mathematical singularity and the initial explosion, by accepting for example that the Universe has some rotation? (In Newton's theory, a rotating star which collapses does not form a singularity unlike the collapse of a non-rotating star.) A lot of effort has been made by mathematicians to answer this question. The most important advances were made in 1969 by *Hawking* and *Penrose*, who showed that *any* model of the Universe which has the observed characteristics of (approximate) homogeneity and isotropy must start from a singularity[5]. This theorem, which does not require absolute homogeneity and isotropy for the model, is one of the most important achievements in the field of relativity. A similar theorem, but not as general, was formulated by *Belinskii, Khalatnikov* and *Lifshitz* in 1970. They discussed some models which, when run backwards in time towards the origin, pass through an infinite number of stages during which they expand and contract in different directions at the same time. All the models lead to a singularity, of the same nature as the singularity of *Hawking* and *Penrose*.

We see, therefore, that the general theory of relativity leads to an initial singularity of the Universe. Would this change if we were not using Einstein's theory but another theory? Several such theories have been developed and we shall discuss some of them in the next chapter. However, so far none of them has managed to replace general relativity. Whenever observational tests were carried out in order to distinguish between relativity and another theory, relativity was vindicated. Consequently, most

---

[5] The Hawking-Penrose theorem holds under the following assumptions: (1) The general theory of relativity holds. (2) The total energy is locally positive (it has recently been shown that this is the case). (3) There are no closed timelike geodesics (i.e. no one can return to his own past). (4) Space is not everywhere flat along all timelike or lightlike geodesics (it is unlikely that this is not the case). (5) There is one closed timelike surface (this can be guaranteed because of the isotropy of the microwave background radiation).

The assumptions, therefore, on which the theorem is based are not very restrictive, and we do not doubt that they apply to the actual Universe. That is why the Hawking-Penrose theorem is very powerful.

researchers today work on relativistic cosmology, rather than on other competing theories.

An effort has been made to avoid the singular beginning of the Universe by introducing quantum mechanical phenomena. Such phenomena are very important when the age of the Universe is less than 10^{-43} sec (Planck time, Sect. 8.7). However, as *Hawking* has pointed out, quantum mechanics does not seem to eliminate singularities. Thus, the basic singularity theorem discussed above cannot be invalidated by quantum mechanics.

The objections, concerning the formation of helium and deuterium, arise from a doubt whether the observed abundances are universal or not. A lot of effort has been put into detecting stars with a helium abundance well below the generally accepted value ($\sim 25\%$). It seems, however, that the "exceptions" observed are not due to reduced helium content but rather to peculiarities in the spectra of certain stars. Besides, the most recent observations with artificial satellites confirm the idea that the observed deuterium has primordial origin, rather than having been formed in more recent stages of the evolution of the Universe.

The objections with respect to the age of the Universe are based on the uncertainties involved in the various calculations. However, it is true that no stars or galaxies have been observed with an age greater than 20 billion years. Other theories, like the theory of continuous creation, claim that there is an infinite number of galaxies older than this limit.

Finally, a lot of effort has been made to try to attribute the microwave background radiation to galaxies or other sources. Recent observations, however, have shown that the isotropy of this radiation is amazing (any irregularities are less than 10^{-4}) and therefore, any non-cosmological origin of it is highly unlikely. Indeed, if this radiation were due to galaxies or stars it would appear more intense in certain directions, contrary to what we observe. It is particularly significant that it has a black-body spectrum, something which would be very unlikely if it were non-cosmological.

The evidence, therefore, for the explosive origin of the Universe is very substantial. So, we shall go on to discuss the first stages of the cosmic expansion, assuming the Big Bang theory to be correct[6].

---

[6] During the 18th General Assembly of the International Astronomical Union (Patras, August 1982) *Zeldovich* gave a talk on modern cosmology and stressed that "the hot Big Bang theory has been established today beyond any reasonable doubt". He compared this theory with the "realisation that the Earth and other planets move around the Sun" [Highlights of Astronomy **6**, 32 (1983)].

6.8 The Distribution of Radiogalaxies

An extra piece of evidence that the Universe was denser in the past comes from the observed distribution of radiogalaxies.

When we observe the distribution of galaxies at larger and larger distances we essentially observe how this distribution was when the light set off from these galaxies. Since the Universe continuously expands, we should see higher densities of galaxies as we observe further in the past. This phenomenon cannot be observed with our optical telescopes because they cannot provide a complete sample of galaxies at very large distances. It can be detected, however, by radio observations. Indeed, *Ryle* (Nobel prize 1974) and his collaborators discovered that the density of galaxies increases with their distance. The reason is that the radiogalaxies are observed to greater distances than optical galaxies, and their number is relatively small, allowing us to explore a complete sample.

If I_0 is the power radiated by a galaxy, the intensity of radiation received by a radio telescope is

$$S = \frac{I_0}{4\pi r^2},\tag{6.38}$$

where r is the distance of the galaxy. If the density of radiogalaxies in space is constant, ϱ, then the number of radiogalaxies out to a distance r is

$$N = \tfrac{4}{3}\pi\varrho r^3\tag{6.39}$$

(assuming space to be Euclidean, which is approximately correct). From the above two formulae, we obtain

$$\log N = -1.5\log S + \log\left[\frac{4}{3}\pi\left(\frac{I_0}{4\pi}\right)^{3/2}\right].\tag{6.40}$$

Therefore, if we plot $\log N$ versus $\log S$ (Fig. 6.6), we expect to get a line with slope -1.5. *Ryle*, however, found a steeper slope, about -1.8, which cannot be explained if the density ϱ and the power I_0 of the radiogalaxies remain constant. It seems, therefore, that either the density ϱ or the power I_0, or very likely both, increase with distance.

This observation was very controversial because it came in direct contradiction to the theory of continuous creation (Sect. 7.3) which required a constant density of radiogalaxies and a constant average value of I_0. Nevertheless, more recent observations confirm Ryle's result. Thus, it really seems that the density of radiogalaxies increases with distance, i.e. increases as we go backwards in time towards the beginning of the Universe. This provides yet more evidence for the fact that the Universe started from a condition much denser than at present.

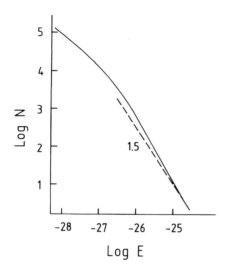

Fig. 6.6. The Ryle phenomenon. The logarithm of the number N of observed radio galaxies versus the logarithm of their intensity. The theoretical curve (---) has slope -1.5, while the observed curve (——) has a steeper slope

6.9 The Age of the Universe

The question of the age of the Universe constitutes one of the major problems of cosmology.

One may pose the same question in a different way: Does the Universe have an infinite past or not? From the information we have available at present, we may answer this by saying that not only is the age of the Universe finite, but that it is between 10 and 20 billion years.

There is a lot of evidence for this. We may use various independent methods to estimate the age of the Universe. The three most important methods are:

a) The expansion of the Universe.
b) The ages of radioactive elements.
c) The ages of the oldest globular clusters.

6.9.1 The Expansion of the Universe

If we assume that all the objects in the Universe have been ejected from a Big Bang, in a Euclidean space, with various velocities which remain constant, their distances from each other will be proportional to their velocities, i.e.

$$r = vt, \tag{6.41}$$

where t is the time elapsed since the initial explosion. We call such an expansion linear. On the other hand, Hubble's law gives $v = Hr$. Therefore, the time from the initial explosion to the present, t_0, is equal to $1/H$. For $H = 50$ km/sec/Mpc, $t_0 = 20$ billion years. This is the so called *Hubble time*.

If the expansion is not linear, the age of the Universe must be less than H^{-1}, because the gravitational attraction decelerates the expansion. This can be seen from the expansion curves for the various models of the Universe (Fig. 6.1). In the case of a closed (spherical) Universe ($\varepsilon = 1$), its age, t_1, is less than $2t_0/3$, i.e. less than 13×10^9 years. The curve $\varepsilon = -1$ represents an open (hyperbolic) model of the Universe with age t_2 between $2t_0/3$ and t_0. Finally, the curve $\varepsilon = 0$ represents a limiting model between the above two cases, with age $2t_0/3$. For $H = 50$ km/sec/Mpc the critical density is approximately 5×10^{-30} g/cm^3. If the Universe has a density equal to the critical density, its age must be exactly $2t_0/3$, i.e. approximately 13 billion years.

If the Universe is open, its age must be more than 13 billion years. *Tammann, Sandage* and others (1980) gave for the deceleration parameter the value $q = 0.02$, which implies that the age of the Universe is 19×10^9 years. However, if the neutrino mass is high enough for the Universe to be closed, its age may be considerably less.

In conclusion, we estimate that the age of the Universe is between 10 and 20 billion years.

6.9.2 The Ages of the Radioactive Elements

The radioactive elements continuously decay, and so cannot live forever. We can use this to estimate the age of the Universe. We must first know the lifetime of the various radioactive elements. It has been estimated that the half-lifes of the two isotopes of uranium U^{238} and U^{235} are 4.5×10^9 and 0.9×10^9 years respectively. The final product of the decay of these isotopes is lead. We get Pb^{206} from U^{238} and Pb^{207} from U^{235}. If we assume that all the existing lead on the Earth came from the decay of uranium, we can find an upper limit for the age of the uranium, which turns out to be 4×10^9 years. Further, from the abundance of the two uranium isotopes U^{235} and U^{238}, we find an age of 6×10^9 years (Sect. 6.10).

Another estimate of the age of the elements is based on the abundance of the isotopes of strontium, Sr, and rubidium, Rb. By this method we estimate the age of the rocks of the Earth. The Rb^{87} decays to Sr^{87} with a half-life of 4.2×10^{10} years. The Sr^{86} does not decay, therefore the ratio Rb^{87}/Sr^{86} reduces continuously, while the ratio Sr^{87}/Sr^{86} increases. Figure 6.7 gives the values of the above ratios for 6 different rocks of the same age. The slope of the curve defines the age of the rubidium and strontium, which is estimated to be 4.54×10^9 years (see Sect. 6.10). Similar measurements have been made with other radioactive elements.

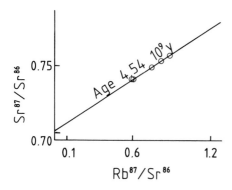

Fig. 6.7. The rubidium to strontium abundance ratio (Rb^{87}/Sr^{86}) versus the abundance ratio of strontium isotopes (Sr^{87}/Sr^{86}) in various rocks. The slope of the curve gives the age of the rocks (the age is the same for all rocks)

In recent years many measurements have been made of meteorites and of lunar rocks. The measurements of meteorites were based upon the relative abundance of potassium (K^{40}) to argon (A^{40}). It was found that the age of the meteorites is 4.6×10^9 years, to very good accuracy. With equally high accuracy, the age of the Moon was estimated to be 4.6×10^9 years.

We note, therefore, that from the radioactivity measurements of the age of the Earth, Moon and meteorites, the age of our solar system is approximately 5 billion years. This number is smaller than the age of our Galaxy, since the Sun, and hence the solar system, belong to the population I stars, which are much younger than the old population II stars. The uranium and other radioactive elements have been formed mainly by population II stars.

Most of the elements were formed during supernova explosions. Their initial relative abundances can be calculated theoretically from the way they were formed. If certain isotopes decay, we may estimate the time that has elapsed from the moment they were formed until today, by comparing their current relative abundance to their initial one.

The elements produced in a supernova explosion are spread throughout the interstellar medium, which consequently becomes enriched with heavy elements such as uranium. Upon formation of the solar system, the interstellar gas, out of which it was formed, ceased to be enriched further; the enrichment stopped 5×10^9 years ago when the interstellar material ceased to be accreted onto the planets and the Sun.

To have an estimate of the age of the Universe we must add the age of the supernovae in our Galaxy to the age of the radioactive elements. The ages of the supernovae, however, are far from certain. Indeed, not all supenovae exploded simultaneously. Today, we observe in our Galaxy, one supernova every 30 to 40 years. During the first stages of the evolution of our Galaxy, however, the supernova rate must have been much higher because there were many more massive stars around.

Further, to find the age of the Universe we must also add the time it took for the galaxies to form, as well as the evolution time for the primordial stars before they reached the supernova stage. Galaxy formation is estimated to last about 1 billion years, while the minimum time for stellar evolution up to the supernova stage was possibly less, given that the more massive stars evolve quite rapidly. However, the time needed for the enrichment of the interstellar medium, out of which the solar system was formed, is rather uncertain. This method eventually yields an estimate of the age of the Universe between 11 and 18 billion years.

6.9.3 The Ages of the Globular Clusters

We can estimate the age of the Universe by finding the ages of the oldest objects in it, i.e. the globular clusters.

We can find the ages of the globular clusters using the Hertzsprung-Russell diagram (Sect. 1.3). Many researchers have estimated the ages of various globular clusters in this way. A systematic study has been done by *Sandage* (1965) who found that their ages range between 11 and 17 billion years.

The oldest clusters of our Galaxy, i.e. the very metal poor clusters M 15 and M 92, were formed 14–17 billion years ago. The metal rich clusters have ages between 11 and 12 billion years. The open (or galactic) clusters are even younger. The oldest of them is NGC 188 which is 6 billion years old.

We may say that the population I and the disc population of our Galaxy, which includes the Sun, is relatively young (5–6 billion years old) while the population II, or halo population, which includes the globular clusters, is 11–17 billion years old.

If we accept that the oldest objects in our Galaxy are 17 billion years old, and add about 1 billion years required for the formation of the Galaxy, we conclude that the age of the Universe is about 18 billion years.

The above method is based upon the current theories of stellar evolution. There is yet another, entirely different method, for estimating the ages of clusters, based upon dynamical considerations.

The stars of a cluster are subject to the gravitational forces of the other stars in the cluster. Thus, they continually change velocity and energy. Stars which acquire velocities in excess of the escape velocity may leave the cluster altogether. The ultimate fate, therefore, of a cluster is its final dissolution. Before this stage is reached, however, a cluster passes through various phases:

1) The initial collapse phase during which the stars are formed, and the mixing of stars of various masses takes place. The duration of this phase is very short, of the same order of magnitude as the "free fall" time of a star from the outskirts of the cluster to its centre. This time is about

10^7 years for a globular cluster. During this stage, the stars of various masses have the same average velocities.

2) The phase of stellar interactions, which leads to a statistical equilibrium. The main characteristic of this equilibrium is the equipartition of energy between stars of various masses. That is, the more massive stars acquire smaller velocities, so that the kinetic energy of each star, $mv^2/2$, is roughly the same independently of its mass. At the same time, the density distribution of the cluster takes a characteristic form known as an "isothermal sphere". This phase lasts much longer than the "free fall" time. The time required for the equipartition of energy, called the "relaxation time", is of the order of 10^9-10^{10} years for a globular cluster.

3) The phase of the formation of a dense core. The more massive stars segregate towards the centre and form a compact core. This stage lasts a few dozen times more than the relaxation time.

4) The phase of the dissolution of the cluster, which requires about 100 relaxation times to be completed. At the end, all that is left of the cluster is a binary or a multiple star system.

From the appearance of a cluster, and more specifically from its density profile, we may decide at which phase of its life a cluster is. The age of the cluster estimated in this way is known as the "dynamical age". Dynamical ages turn out to be 10–20 billion years. This method is not very accurate, but it shows that dynamical considerations give ages of the same order of magnitude as methods based upon stellar evolution.

Comparing the results of the various methods for estimating the age of the Universe, we see that they all fall between 10 and 20 billion years, in spite of the fact that all the methods are independent of each other. This very significant agreement in the results indicates that all the phenomena in the Universe had a common origin. This is one of the most important conclusions of modern cosmology.

6.10 The Ages of the Chemical Elements

It is known that if there were N_0 radioactive nuclei at time $t = 0$, the number of nuclei left after time t is given by the law of radioactive decay:

$$N = N_0 e^{-\lambda t}, \tag{6.42}$$

where λ is a constant, independent of the state of the radioactive material, its temperature etc. The half-life is the time after which only half of the initial number of nuclei is left, and is given by

$$t_{1/2} = \frac{\ln 2}{\lambda}. \tag{6.43}$$

Assuming that all the lead (Pb) we observe today on the Earth came from the uranium decay (U_0), we can find an upper limit for the age of the uranium. The amount of lead today is given by

$$Pb = U_0 (1 - e^{-\lambda t}), \tag{6.44}$$

while the amount of uranium today is

$$U = U_0 e^{-\lambda t}. \tag{6.45}$$

Combining (6.44) and (6.45) we obtain

$$\frac{Pb}{U} = e^{\lambda t} - 1. \tag{6.46}$$

From the observed relative abundance of lead and uranium (Pb/U) we find that $t \simeq 4 \times 10^9$ years.

Similar formulae hold for the relative abundance of rubidium/strontium. Rubidium decays into strontium 87, while strontium 86 is stable. Suppose that initially we had Rb_0^{87}, Sr_0^{87} and Sr_0^{86} nuclei. The corresponding numbers of these nuclei today are:

$$Rb^{87} = Rb_0^{87} e^{-\lambda t}, \tag{6.47}$$

$$Sr^{87} = Sr_0^{87} + Rb_0^{87} (1 - e^{-\lambda t}) \quad \text{and} \tag{6.48}$$

$$Sr^{86} = Sr_0^{86}, \tag{6.49}$$

because Sr^{86} does not decay. We have, therefore,

$$\frac{Sr^{87}}{Sr^{86}} = \frac{Sr_0^{87}}{Sr_0^{86}} + \frac{Rb^{87}}{Sr^{86}} (e^{\lambda t} - 1). \tag{6.50}$$

Different rocks which have the same age have different ratios Sr^{87}/Sr^{86} and Rb^{87}/Sr^{86}, but the same ratio Sr_0^{87}/Sr_0^{86} (the same isotopic composition of the original strontium). Therefore various rocks of the same age, t, are represented by points on the same straight line in a coordinate system Rb^{87}/Sr^{86} versus Sr^{87}/Sr^{86}. The age is found from the slope of the line, which is equal to $(e^{\lambda t} - 1)$, (see Fig. 6.7). With this method we find that the age of the Earth is about 4.6×10^9 years.

Another way of finding the age of the chemical elements is based on the relative abundance of the uranium isotopes U^{238}/U^{235}. Both of them decay, but U^{235} decays much faster than U^{238}. Using (6.45) we find

$$\frac{U^{235}}{U^{238}} = \frac{U_0^{235}}{U_0^{238}} \frac{e^{-\lambda_1 t}}{e^{-\lambda_2 t}}, \tag{6.51}$$

where λ_1 and λ_2 are the decay constants of the two isotopes respectively. If the quantities of the two isotopes were initially the same, we find that the age of the uranium is roughly 6×10^9 years.

6.11 Different Models Derived from the General Theory of Relativity

The models of the Universe based on general relativity which we have studied so far are the simplest ones. They are the Friedmann-Lemaitre models with cosmological constant $\lambda = 0$. There is, however, an infinite number of more general models, many of which have been studied in recent years. The simplest of them are the homogeneous and isotropic ones with cosmological constant different from zero. Next in complexity are the anisotropic but homogeneous models, and finally the anisotropic and inhomogeneous models.

Hawking and *Penrose* studied the most general anisotropic and inhomogeneous models (1969). As we mentioned in Sect. 6.7 all anisotropic and inhomogeneous models, which are compatible with the observed characteristics of the Universe, lead to an initial singularity, an "initial explosion". The remaining problem to be discussed is the behaviour of the Universe close to the initial singularity. It is usually assumed that the Universe is asymptotically homogeneous as we approach its origin.

Another extreme idea has been suggested by *Misner* (1969). His model is called "chaotic" because close to the origin of the Universe it passes through an infinite number of successive states of different form, which can be characterised as chaos. This theory assumes that various phenomena during the early Universe made it homogeneous and isotropic. Such phenomena inlcude the formation of pairs of particles during the Planck era, i.e. when the age of the Universe was less than 10^{-43} sec (Sect. 8.7).

However, we have no observational evidence to suggest that there was an initial chaotic phase. The primoridal perturbations in the expanding Universe which formed the galaxies, are much less than the irregularities one might expect in a completely chaotic Universe.

Therefore some cosmological solutions of Einstein's equations may not apply to the real Universe.

7. Other Cosmological Theories

7.1 Newtonian Cosmology

Besides the general theory of relativity, several other cosmological theories have been proposed. Some of them are fully formulated with specific predictions which can be checked, while others are simply vague ideas.

It is not possible here to expand upon all the cosmological theories which have been proposed at various times. We shall discuss only the most important of them. The simplest of all is Newtonian cosmology.

Milne and *McCrea* (1934) have shown that all models of the expanding Universe, which are homogeneous and isotropic solutions of Einstein's field equations, can be derived from Newtonian mechanics as well. The only requirement is that besides the gravitational attraction which tends to bring the matter in the Universe together, we must also have an "initial" expansion velocity, i.e. a Big Bang. Furthermore, if the cosmological constant λ is not zero, it can be interpreted as a repulsive force proportional to the distance, which will also contribute to the expansion of the Universe. If the density of the Universe is high enough for the expansion to be followed by a contraction, the Universe will pulsate. If, however, the density of the Universe is too low for this, its expansion will continue forever.

It is clear that this theory does not come to any different conclusions from those of general relativity in the case of homogeneous and isotropic models of the Universe. However, the two theories do not agree in the general case of anisotropic and inhomogeneous models. Thus, we rely more on general relativity because we know that it is more accurate than Newtonian theory. For this reason, Newtonian cosmology can be considered as only an interesting exercise.

7.2 Kinematic Relativity

This theory was formulated in 1935 by *Milne*. It is based on an original definition of space and time. According to *Milne,* only the concept of time is fundamental. The distance between two points *A* and *B* is *defined* by the time it takes light to travel between them, multiplied by the speed of light (radar method). The new element in this theory is that the clocks which measure time can be arbitrary. We can have, therefore, an infinite number of different times. *Milne* distinguishes two basic types of time, the dynamical time τ and the kinematic time t. With respect to the first one, the galaxies are stationary and the Universe does not expand. On the other hand, the relative velocity of particles separated by a distance r, with respect to the second time, is

$$v = \frac{r}{t}. \tag{7.1}$$

This formula coincides with Hubble's formula for small distances, provided we put

$$H = \frac{1}{t}, \tag{7.2}$$

where t is the "age of the Universe" (Sect. 6.9.1).

The relation between t and τ, according to *Milne*, is

$$\tau = t_0 \log \frac{t}{t_0} + t_0, \tag{7.3}$$

where t_0 is the present time. By this definition, at present the times τ and t coincide. The above formula, however, gives $\tau \to -\infty$ as we approach the origin of time $t = 0$, i.e. the past of the Universe is finite in t but infinite in τ. This theory further accepts that the kinematic time is measured by the atomic phenomena (atomic time) while the dynamical time is measured by the motions of the planets (Newtonian time[1]).

This distinction between the two times is interesting. In the past few years special effort has been made to check whether there are any differences between atomic and Newtonian time, using artificial satellites.

Milne's theory has been essentially abandoned today. It is mentioned here mainly because it was the first theory to use the concept of two different times in cosmology.

7.3 The Theory of Continuous Creation

This theory was developed in 1948 in two closely related versions, by *Hoyle* and by *Bondi* and *Gold*.

---

[1] "Newtonian" time in astronomy is the time *defined* by the motions of the planets, with relativistic corrections taken into account.

According to the theory of continuous creation, the Universe looks approximately the same to all observers, no matter where they are or when they live. This is the "perfect cosmological principle" which states that the Universe is not only isotropic and homogeneous, but it also appears the same at all times. This is why the theory is also known as the "steady state theory". The immediate implication is that the density of the Universe is constant in time. Therefore, since the Universe expands and the galaxies move away from each other, new matter must be created *out of nothing* to replenish the mass which moves out of a certain region. (Creation *out of nothing* does not mean creation of matter out of energy, because matter and energy are considered one and the same thing.)

The amount of matter created cannot be detected experimentally as it is very small; only one hydrogen atom is created per litre per 10^9 years. In the vast span of space, however, the matter created at this rate is enough to produce new galaxies and maintain the average density of the Universe constant. This continuous creation of matter is not explained, but simply accepted as a new principle. The theory implies that the age of the Universe is infinite, although every object in it has a finite age.

The most important contribution this theory made to cosmology is the interest it stimulated for research in the major cosmological and astrophysical problems. For example, it encouraged research on stellar nucleosynthesis, on the large scale distribution of galaxies, and so on.

The weak points of this theory are the following:

1) It cannot account for the origin of the microwave background radiation. All attempts to attribute this radiation to galaxies or other sources in the Universe have failed. The only plausible explanation for the origin of this radiation remains the proposition that the Universe passed through a state of very high density and temperature. This explanation is in complete contradiction with the "perfect cosmological principle" and, therefore, with the theory of continuous creation.

2) The helium production in stellar interiors is inadequate to account for the observed proportion of this element in the Universe. The currently accepted theory for the production of helium is the α-β-γ theory proposed by *Gamow* and his collaborators. This theory accepts that most of the helium was created soon after the Big Bang.

3) According to the theory of continuous creation, the deceleration parameter q is -1. As we saw in Sect. 6.6, although the observations are not accurate enough for an exact determination of q, they favour positive values of q, and rather exclude the value $q = -1$.

4) If the theory of continuous creation is right, we should see the same density of galaxies no matter how far we may observe. This is because observations of more remote regions of the Universe reveal how the Universe used to be as long ago as it took the light from those regions to reach us. In a steady state Universe, no evolution with time is expected,

and, therefore, even the distant regions should have the same density of galaxies as in our local vicinity, which we see as it is at the *present* time. This is contrary to the observations of radiogalaxies which seem to be more densely concentrated at large distances (Sect. 6.8).

The above reasons led most of the supporters of the theory of continuous creation to abandon it.

Recently, *Hoyle* (1980) proposed a new version of the theory. He suggested that the new matter comes from "white holes" with sizes equal to galaxies. His major argument is that the evolution of the present forms of life starting from inorganic materials, is expected to take much longer than the Hubble time ($t_0 = H^{-1}$), i.e. the usually quoted "age of the Universe". This argument, however, is very weak because the biological phenomena are very poorly understood. This is one reason why these new ideas have not been received very favourably by the astronomical community.

7.4 The Brans-Dicke Theory

The *Brans-Dicke* theory (1961) applies Mach's principle to cosmology in an original way. According to Mach's principle, the inertia of every particle (and therefore its mass m) depends upon the influence of the rest of the Universe. *Brans* and *Dicke,* therefore, concluded that since the expansion of the Universe dilutes the matter in it, this influence must decrease with time. A quantity which measures the influence of the matter in the Universe on a particle of mass m is $m \sqrt{G/(hc)}$ where h is Planck's constant and G the gravitational constant. We have good reasons to believe that Planck's constant and the speed of light, c, remain absolutely constant in time. Then any change in the above quantity must be due to either a change in G, or to a change in the mass m. So, there are two versions of the Brans-Dicke theory: one which considers that the masses of the particles remain constant, while the gravitational constant slowly and continuously decreases; and another which considers that G is constant, but the masses of all particles in the Universe change. These two versions are related by a transformation, so that they can really be considered as one theory.

The divergence of the Brans-Dicke theory from general relativity is characterised by a parameter ω. For small values of this parameter, the differences between these two theories are large, while for large values of ω the differences between the two theories tend to disappear.

The Brans-Dicke theory may be checked against observations in our solar system. For example, the theory predicts that the perihelion shift in

Mercury's orbit is Δ, which is related to the prediction of general relativity Δ_{GR} by

$$\Delta = \frac{3\omega + 4}{3\omega + 6} \Delta_{GR}. \tag{7.4}$$

The observations agree very well with the relativistic prediction. *Dicke*, however, argues that part of the shift in Mercury's perihelion is due to the oblateness of the Sun. So he calculated that the true value of Δ is 94% of Δ_{GR} and therefore $\omega = 6$. More recent observations, however, show that the oblateness of the Sun is insignificant and that ω is greater than 30.

Another test for this theory is the divergence, D, of the light rays which pass close to the Sun. The Brans-Dicke theory predicts that

$$D = \frac{2\omega + 3}{2\omega + 4} D_{GR} \tag{7.5}$$

where D_{GR} is the relativistic prediction. As we saw in Sect. 5.4 the observations agree very well with the relativistic prediction. The observational errors allow us only to say that $\omega \geq 30$.

If the value of ω is as low as 6, the corresponding change in G is very large, and this has important consequences for the evolution of stars. This is because G enters to the seventh power (G^7) in the crucial quantities affecting stellar evolution. Various studies of stellar evolution with varying G have been made, and the comparisons with observations show that ω must be much higher than 6. Thus the differences between the Brans-Dicke theory and general relativity are insignificant.

Further, the cosmological predictions of the Brans-Dicke theory are qualitatively the same as the predictions of general relativity. There are quantitative differences, however, the main being that this theory predicts a faster expansion rate in the early Universe. The Brans-Dicke theory also predicts an initial Big Bang with all the consequences mentioned in the previous chapter. Given that even if the theory is correct, its differences from relativity are small, interest in it is limited.

7.5 Dirac's Theory

Dirac's theory (1937, 1973) is based upon the observation that there are some very "large numbers" in the Universe which are approximately equal to one another. These numbers are:

a) The ratio of the electromagnetic force between a proton and an electron to the gravitational force between them:

$$\frac{e^2}{G m_p m_e} \simeq 10^{40}, \tag{7.6}$$

where e is the charge of an electron and m_p, m_e are the masses of the proton and electron, respectively.

b) The age of the Universe T ($10-20 \times 10^9$ years) in units of the "elementary time" $t_e = r_e/c$ [defined as the time required for light to cross an "electron radius" $r_e = e^2/(m_e c^2)$]. The ratio T/t_e is about 10^{40}.

c) The square root of the total number of particles in the Universe. (In the case when the Universe is infinite, we consider the number of particles inside a sphere of radius $c t$, where t is the age of the Universe, that is, we consider the "observable" Universe.) The total number of particles in the Universe, defined in this way, is $N \simeq 10^{78}$, and its square root is: $\sqrt{N} \simeq 10^{39}$, i.e. very close to 10^{40}.

Dirac suggested that the coincidence of these three numbers, of such different nature, is not fortuitous but must express a fundamental law of nature. So, he proposed that the above three numbers are necessarily equal.

The second of these numbers, however, changes with time, as the age of the Universe increases. Therefore, the other two numbers must change as well, i.e. the quantities $e^2/(G m_p m_e)$ and N increase continuously with time.

If we accept that e, m_p and m_e are constant, G must vary as

$$G \propto \frac{1}{t},$$ (7.7)

that is, G is inversely proportional to the age of the Universe. Also, the number N of particles in the Universe must increase as the square of the age of the Universe:

$$N \propto t^2.$$ (7.8)

Therefore, according to this theory, new matter is continuously being created. The creation of matter may be either (a) uniform everywhere in the Universe (additive process) or (b) close to the already existing matter (multiplicative process).

Dirac's theory, which is also known as the "large number theory", has similarities with the Brans-Dicke theory, on the one hand, and the theory of continuous creation on the other. However, *Dirac* does not accept that the total mass of the Universe increases. For this reason, in the multiplicative process he introduces a further assumption that the mass m of each particle reduces so that the total mass $N m$ remains constant; in the case of the additive process, equal quantities of positive and negative masses are produced (with the mass of each particle remaining constant) so that the total mass of the Universe does not change. The negative mass, however, cannot be observed.

Finally, *Dirac,* just like *Milne,* introduces two times, the dynamical time τ (or Einstein time, as he calls it) and the atomic time t. The relation between these two times is the same as *Milne's* relation (7.3).

In a more recent version (1979), *Dirac* makes an effort to reconcile his theory with general relativity. *Dirac* now accepts that the mass of each particle is constant, so G varies according to (7.7), and the two times, the Einstein and atomic time, have a common origin, being related by

$$\tau = \frac{t^2}{2}. \tag{7.9}$$

This new version of the theory, therefore, introduces a Big Bang, as does relativity. The age of the Universe, however, is found to be 6×10^9 years, i.e. too low. That is why this theory is considered rather implausible.

The most important test of this theory is the measurement of the variation in G. Dirac's theory predicts an even larger variation in G than the Brans-Dicke theory. Therefore, the same arguments given in Sect. 7.4 hold against both theories.

Even so, the experimental efforts to measure any variation in G with time are continued. One of the recent estimates (1981) is that the relative variation of G is

$$\frac{\dot{G}}{G} = 6 \times 10^{-11}/\text{year}, \tag{7.10}$$

where $\dot{G} = dG/dt$ is the change of G per unit time. The possible implications of such a variation are still being studied.

Dirac's theory does not explain *why* the large numbers we come across in the Universe are approximately equal. *Eddington* was the first to attempt such an explanation. His explanation, however, was unsuccessful.

On the other hand *Carter* (1974) gave an entirely different explanation for the coincidence of the large numbers. *Carter* noticed that if those numbers did not approximately coincide, life and man would not exist in the Universe. Therefore, the coincidence of Dirac's large numbers is a necessary prerequisite for man to exist in the Universe. This is the "anthropic principle" that will be discussed in Sect. 11.5.

7.6 Jordan's Theory

There are many theories which are similar, in one way or another, to the theories we have already discussed. One of them has been developed by *Jordan* (1949). *Jordan* noticed that if we add the kinetic energy of the expansion of the Universe, the rest mass energy of the Universe and its potential energy (which is negative), we find zero, to a rather high accuracy:

$$E = \sum \frac{mv^2}{2} + \sum mc^2 - \sum\sum \frac{Gmm'}{r} = 0. \qquad (7.11)$$

(The summations $\sum$ refer to all particles in the Universe m and m' and r is the distance between two particles. The Universe is assumed to be finite.)

This observation led *Jordan* to introduce the assumption that the sum of the mass-energy in the Universe is *always* zero. Since the Universe expands, however, the potential energy term becomes smaller. For the total mass-energy of the Universe to remain zero, *Jordan* then assumes that new matter is created in the Universe, at appropriate distances and with appropriate velocities.

This theory includes the interesting "philosophical" point of view, that the total matter-energy of the Universe is zero, therefore what has been created out of nothing, is nothing. Even the "initial creation", according to *Jordan*, consisted of the creation of two hydrogen atoms at such distance from each other and with such velocities, that their total mass-energy was zero. We may say, therefore, that the Universe is something even less than a huge soap bubble, because if we condense it, we will simply get nothing out of it.

Jordan's theory cannot explain the microwave background radiation and other observational data. That is why it is not considered seriously today. His basic idea, however, that the total mass-energy of the Universe is zero, was revived recently by *Zeldovich* and others.

7.7 Conclusions

We have examined in this chapter various theories which have been put forward at various times, hoping to replace Einstein's theory of general relativity. There are other, similar theories to those we mentioned, of lesser importance.

Possibly the most important of these alternative theories has been the theory of continuous creation, not only because many people worked on it, but also because it attempted to replace the origin of the Universe by an entirely different idea, the continuous creation of matter out of nothing. Thus this theory generated a great "philosophical" interest as an idea opposed to the idea of an initial creation.

We notice that every cosmological theory incorporates certain interesting, and sometimes revolutionary ideas. In general, however, the various theories are short-lived. Many theories survive only as long as it takes to devise a test, a way of checking the theory, which brings to light its weak points. Then the theory is either rejected immediately, or its supporters try to extend its life by modifying it, until a new test is devised to reject the new version. The theory of continuous creation followed such a course. In its original form, the new matter in the Universe was believed to be created in the form of hydrogen atoms, while the heavier elements were believed to be formed later, in the stellar interiors. When it became clear that the theory could not explain the observed helium abundance in the Universe, some of its supporters added the assumption that helium is also created out of nothing. If, however, we accept this, why shouldn't we accept that other elements are also created out of nothing? In this way, the theory loses its initial, attractive simplicity. If we are ready to accept a particularly complicated theory, why shouldn't we accept that the Universe was created as it is today, 8000 years ago, according to the ideas of the "fundamentalists"? The answer is that we always try to find the simplest possible theory which, with the minimum number of assumptions, can explain the present form of the Universe.

In cosmology, just as in phyiscs, the principle of "Occam's razor" holds. According to this principle, no assumptions should be made apart from those which are absolutely necessary. Furthermore the criterion by which we judge the value of a new idea, or assumption, is how well it stands the tests, both theoretical (many theories are rejected as lacking self-consistency) and observational.

From this aspect, general relativity has shown an amazing longevity. The hostility it initially provoked from all directions benefitted it, because it showed its endurance. The various experiments devised to distinguish between this theory and other competing theories always supported general relativity. In no case was a modification of its original version necessary. That is why we particularly trust it, and more studies are still being carried out to check its implications.

It is possible that general relativity did not hold when the age of the Universe was less than the "Planck time" (10^{-43} sec). This is when the radius of the visible Universe of today was about 10^{-4} cm, and its temperature was greater than 10^{32} K. The related theoretical studies are known as "quantum gravity", and we shall discuss them in Sect. 8.7.

As yet, however, there is no generally accepted theory of quantum gravity. Even if one day we have a theory more accurate than relativity, which will describe the first 10^{-43} sec of the Universe, Einstein's theory will still have accomplished a very important feat. The fact that it dominated physics for so many years, and explained so many phenomena concerning the matter in the Universe. In this respect, relativity can be compared to Newtonian theory, which dominated astronomy for centuries. The fact that relativity replaced Newton's theory, does not mean that Newton's theory is useless. We simply know where it applies, and where it does not apply. Relativistic corrections are, in most cases, insignificant and only in extreme cases, such as black holes, are Newton's laws not applicable at all. We may say that Newton's theory proved to be much more applicable than one would have expected from the experimental evidence Newton had. Similarly, Einstein's theory has many more applications today than its first advocates could have imagined. It is plausible that one day we shall find its limitations. Until then, however, we shall use it in the problems we have to solve.

8. The Beginning of the Universe

8.1 The First Stages of the Expansion

As we saw, the currently accepted theory for the evolution of the Universe is the theory of the Big Bang. This is the so-called "standard theory". The first stages of the expansion within the framework of this theory are particularly interesting. They are concerned with the structure of the Universe a long time before the formation of galaxies and stars, when the cosmic temperature was greater than 3000 K. Even more significant were the first few minutes of the Universe, during which the elementary particles and the first chemical elements were formed. *Weinberg* (Nobel prize 1979) has written an interesting book on this subject, called "The First Three Minutes" (1977). As we shall see, the physics of the first stages of the Universe is not a matter of speculation any more, but rather a subject of serious scientific research.

In Sect. 8.2 we explain why, during the first stages of the expansion, the radiation energy was greater than the matter energy. Today the radiation density is less than 1/1000 of the matter density in the Universe. When, however, the temperature was approximately equal to 4000 K and the age of the Universe was 500 000 years, the radiation density was equal to the matter density. Before that time, the radiation dominated over the matter. This initial stage of the Universe is called the "radiation era", while the subsequent one is the "matter era" (Fig. 8.1).

Soon after the beginning of the matter era, when the age of the Universe was 700 000 years, another significant event happened: the creation of atoms out of nuclei and electrons. The temperature of the Universe, at this stage, was 3000 K. At greater temperatures the atoms were ionised because of continuous collisions between particles and photons. As the temperature dropped below 3000 K the photons did not have enough energy any more to ionise the matter entirely, so neutral atoms could be formed. Most of the electrons took part in the formation of neutral atoms, so that only a few free electrons were left afterwards, and the interaction between photons and electrons ceased. At the same time the mean free path of the photons, due to scattering by neutral atoms, became as large as the size of the horizon. Therefore, the photons became decoupled from the matter and were free to move throughout space. Today, these photons make up the "microwave background radiation".

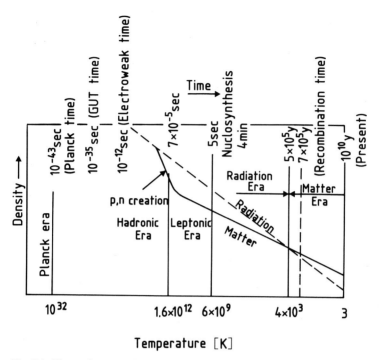

Fig. 8.1. The various eras in the early Universe

As the Universe expanded, the wavelength of the photons expanded as well, at the same rate as the distance between the galaxies. The frequency of the photons, therefore, reduced and so did their energy. The temperature which corresponded to this photon energy also reduced at the same rate (Sect. 8.2), and the temperature today is approximately 3 K, corresponding to a photon wavelength of a few centimeters or millimeters. We may say, therefore, that the microwave radiation set off at the time of the formation of atoms. This time is called the "recombination time" ($t_{rec} = 700\,000$ years).

Before the recombination time, the various particles interacted strongly with each other. However three basic quantities were conserved, namely: (a) the electric charge, (b) the baryon number and (c) the lepton number. The baryon number is equal to the number of baryons (i.e. protons, neutrons, hyperons etc.) minus the number of antibaryons. The lepton numer consists of three separate numbers, the electron lepton number, the muon lepton number and the tau lepton number. These three numbers correspond to the electrons, muons and the heavy (tau) leptons, as well as their corresponding neutrinos.

The electric charge, the baryon number and the lepton number of the Universe seem to be small. For example, the total electric charge of the Universe must be zero, or very close to zero. That is, the Universe contains the same number of negative and positive charges. This may be inferred from the current observations that the matter in the Universe is on the average neutral today, and the assumption that the basic constants of the Universe remain unchanged with time. For example, if the Earth were positively charged, even as slightly as having one extra positive charge per 10^{36} charges, the electric forces between the Earth and the Sun would dominate gravity, which is not the case. Therefore, the excess of positive (or negative) charges is either zero or very close to zero.

The baryon number is relatively small, that is, there is one baryon per 10^9 photons. Nevertheless, as we shall see later on, during the first moments of the Universe, the creation of baryons and antibaryons out of photons was very efficient. Most of them were annihilated with each other, and only a small percentage survived.

Some people have expressed the point of view that the baryon number is precisely zero, and therefore the Universe is made up of equal quantities of matter and antimatter. However, as we shall see in Sect. 8.6 this does not seem very likely, as we do not have any indication that there is, elsewhere in the Universe, a large quantity of antimatter. More specifically, the cosmic rays which come from our Galaxy, as well as from other galaxies, do not seem to contain significant quantities of antimatter.

The lepton number is also considered to be small. Charge conservation and the lack of antimatter requires equal numbers of protons and electrons. During the first stages of the Universe, there were many more electrons and positrons around, but they annihilated with each other, just like the protons and antiprotons. Thus only a small number of electrons was left. At the same time a large number of neutrinos and antineutrinos was created from reactions such as

$$e^- + e^+ \rightarrow v_e + \bar{v}_e.$$

These neutrinos still exist in the Universe. The lepton number, however, depends upon the difference between the numbers of neutrinos and antineutrinos, and is most likely small. In any case, our conclusions do not depend strongly upon the lepton number, so our uncertainty about its value does not really matter.

As we go backwards in time, we see that the temperature of the Universe increases significantly, and when $t \rightarrow 0$ the temperature tends to infinity. When the temperature exceeds a certain limit, the energy of the photons is so high that various elementary particles can be created out of them, in particle-antiparticle pairs. The required energy is $2mc^2$, where m is the rest mass of each particle; as a function of temperature T, this energy is kT, where k is Boltzmann's constant. Therefore, for a certain type of particle of mass m to be produced, the temperature of the Universe must be

$$T \geq 2mc^2/k. \tag{8.1}$$

Table 8.1. Minimum temperatures for the formation of various particles

Particles	Critical temperature	Corresponding age of the Universe
Leptons		
Electrons and positrons (e^-, e^+)	6×10^9 K	5 sec
Muons (μ^-, μ^+ and their antiparticles)	1.2×10^{12} K	1.2×10^{-4} sec
Hadrons [a]		
Mesons $\pi(\pi^0, \pi^+, \pi^-$ and their antiparticles)	1.6×10^{12} K	7×10^{-5} sec
Nucleons (protons, antiprotons, neutrons, antineutrons p, $\bar{p}$, n, $\bar{n}$)	10^{13} K	1.5×10^{-6} sec

[a] Hadrons are the baryons (nucleons, hyperons, etc.) and the mesons. All these particles interact with the strong force (Sect. 8.5)

The minimum temperatures required for the production of various particles are approximately given in Table 8.1.

We notice that plenty of electrons and positrons were created when the temperature was greater than 6×10^9 K, while when the temperature was greater than 1.6×10^{12} K a great number of mesons and antimesons were formed. Nucleons (protons and neutrons) as well as their antiparticles were abundant at temperatures greater than 10^{13} K.

The period between 7×10^{-5} sec (temperature 1.6×10^{12} K) and 5 sec (temperature 6×10^9 K) is called the "lepton era", while the period before 7×10^{-5} sec is called the "hadron era". We distinguish the very early era, before $t = 10^{-43}$ sec (temperature $> 10^{32}$ K) and call it the "Planck era" (Fig. 8.1). During this very early period gravity was probably quantized and unified with the other forces of the Universe (Sect. 8.7).

At first sight there appears to have been much more matter around during the hadron and lepton eras than during the "matter era", which came later on (Fig. 8.1). The behaviour of the matter during these early epochs, however, was very different from its usual behaviour. When the temperature was much higher than mc^2/k the particles had much more energy than their rest energy mc^2. Therefore, their rest energy could be neglected, that is the particles behaved like photons which have zero rest mass. In general, the matter had all the properties of a form of radiation and more specifically, the energy of the matter was proportional to the fourth power of the temperature, just as for photons (while during the "matter era", the energy of the matter is proportional to the third power of the temperature; Sect. 8.2). During the first stages of the Universe, therefore, the distinction between matter and radiation was much less clear.

After these general comments, let us see in more detail some "pictures" of the expanding Universe at various times, summarised from *Weinberg*. These pictures represent the major characteristics of the Universe as its temperature gradually reduced from 10^{11} K to 10^9 K.

8.1.1 Temperature 10^{11} K (Age of the Universe $t = 0.01$ sec)

There is thermal equilibrium between photons and particles. Electrons, neutrinos and their antiparticles (positrons and antineutrinos) are being created in abundance. The creation of hadrons has ceased, since this requires much higher temperatures. The protons and neutrons are being continuously converted into each other. In this way they are in more or less equal proportions, 50% protons and 50% neutrons. No complex chemical elements can be formed, as any concentration of protons and neutrons is immediately dispersed.

8.1.2 Temperature 10^{10} K (Age of the Universe $t = 1.1$ sec)

The weak interactions drop out of thermal equilibrium. Hence neutrinos and antineutrinos are no longer in thermal equilibrium with the other particles. Instead they move freely in space. Their energy reduces continuously due to the cosmic expansion. According to the theoretical predictions, if we could observe these cosmological neutrinos, which fill space uniformly, they should have a temperature of 2 K today. At present there is no known way to observe this radiation, because the neutrinos interact very weakly with matter. (To stop a neutrino we need, on average, a wall of lead many light years thick.) However, as the number of these neutrinos is enormous (it is estimated that their total energy is at least 45% of the total energy of the photons), it is not out of the question that one day this neutrino radiation may be detected. Such a discovery will be one of the most important advances of cosmology.

At the temperature of 10^{10} K there are more protons than neutrons (76% protons and 24% neutrons). This is due to the fact that the mass of the neutron is slightly larger than the mass of the proton, and the two species are kept in equilibrium by the weak interactions. The reactions producing the protons are thus favoured over the reverse reactions which produce the neutrons. When the weak interactions drop out of equilibrium, the ratio of protons to neutrons "freezes" at the value reached at that temperature.

8.1.3 Temperature 10^9 K (Age of the Universe $t = 3$ min)

This temperature is well below the minimum temperature required for the creation of electron-positron pairs. The photons have, therefore, ceased forming such pairs and most of the already formed electrons and positrons have annihilated with each other. Only some electrons have been left, about equal in number to the protons, so that the total charge of the Universe is zero, or almost zero.

At this stage the light elements are formed in great quantities, mainly deuterium D (or heavy hydrogen H^2) and helium ($_2He^4$). The collisions between protons and neutrons (which are now in the proportion 86% protons to 14% neutrons) form deuterium nuclei. Subsequently deuterium collides with protons and neutrons and forms helium (Fig. 6.5). All the deuterium and helium is formed in a very short time interval. At the end of this stage the temperature is $T = 0.9 \times 10^9$ K and the age of the Universe is $t = 3$ min 45 sec. A little later, practically all the neutrons have been incorporated into helium $_2\text{He}^4$ and other elements, so that no more neutrons are left for nucleosynthesis to continue. Only much later are heavier elements formed in stellar interiors.

Cosmological nucleosynthesis lasts for approximately 4 minutes (from $t = 10$ sec to $t = 3$ min 45 sec). This so important stage of the cosmic evolution had already been described by *Gamow* and his collaborators by the year 1948 (Sect. 6.7).

Between the end of nucleosynthesis ($t = 4$ min) up to the epoch of the formation of the atoms ($t_{\text{rec}} = 700\,000$ years) nothing important happened. After the recombination time the most important event in the Universe was the formation of galaxies and stars, which we shall discuss in Sect. 8.3.

8.2 Energy of Radiation and Matter

The wavelength λ of a certain photon and the scale factor R of the Universe are changed by the same factor f by the expansion of the Universe. That is, if R becomes f times larger, $R' = fR$, then λ becomes also f times larger:

$$\frac{\lambda'}{\lambda} = \frac{R'}{R} = f. \tag{8.2}$$

The ratio f is called the "expansion factor". The implication is that the frequency v of a photon, as well as its energy hv, reduce by the same factor f.

On the other hand, the number of photons inside a certain volume reduces by a factor f^3. Therefore, the total energy density of the radiation reduces by a factor f^4.

The radiation density, however, is proportional to the fourth power of the temperature (Stefan-Boltzmann law). Therefore, the temperature reduces by a factor f, that is

$$\frac{T'}{T} = \frac{R}{R'} = \frac{1}{f}. \tag{8.3}$$

The density of the matter reduces by a factor f^3, because the mass-energy of each particle does not change, assuming that the largest part of the energy in the matter is the rest mass energy ($E = mc^2$).

Since the radiation energy is proportional to R^{-4} while the matter energy is proportional to R^{-3}, for small enough R the radiation energy exceeds the matter energy (Fig. 8.1). The epoch during which the radiation had more energy than the matter (radiation era) finished when the age of the Universe was approximately $t = 500\,000$ years, and its temperature was 4000 K.

The relation between the scale factor R and the age of the Universe was different in the radiation era and in the matter era. During the radiation era the dimensions of the Universe increased according to the law

$$R \propto t^{1/2}, \tag{8.4}$$

while during the matter era, R increases according to the law

$$R \propto t^{2/3}. \tag{8.5}$$

We notice that as t tends to zero, (8.4) implies that R tends to zero abruptly. The rate of change of R tends to infinity as t tends to zero. In fact:

$$\frac{dR}{dt} \propto t^{-1/2}, \tag{8.6}$$

and $dR/dt \to \infty$ as $t \to 0$.

8.3 The Formation of Galaxies and Clusters of Galaxies

The most important event after the formation of atoms in the early Universe was the formation of galaxies, and of clusters (or superclusters), as well as the formation of the first stars.

The formation of galaxies and clusters of galaxies consists of two stages:

a) The formation of the initial density concentrations (protogalaxies and protoclusters) out of small fluctuations of density in the initial stages of the expansion of the Universe, and

b) the collapse of the protogalaxies to form the galaxies we observe today.

The first phase started well before the epoch at which the atoms formed ($t_{\text{rec}} = 700\,000$ years) and perhaps during the Planck era ($t_p = 10^{-43}$ sec). The formation of galaxies was essentially complete by the time the age of the Universe was $t = 10^9$ years or even earlier, depending upon which theory of galaxy formation is correct. On the other hand, the collapse which led to the present forms of galaxies was relatively fast (taking roughly 3×10^8 years). During this collapse, the galaxies were fragmented into stars. Indeed, the oldest stars of the galaxies (population II stars) were formed during the collapse of the galaxies. Subsequently, when a dense layer of interstellar medium had been accumulated in the plane of symmetry of the spiral galaxies, and particularly when the spiral arms were formed, local concentrations of matter formed population I stars, i.e. stars like the Sun. Stars of this kind are continually being created in spiral galaxies, and in many irregular galaxies.

Some years ago it was suggested (*Rees* 1977) that there is yet another stellar population (population III) which was formed before the galaxies, that is during the protogalaxy stage. This population is postulated to explain the presence of a small percentage of "metals" even in the oldest stars of population II.

8.3.1 Theories of the Formation of Protogalaxies and Protoclusters

Two mechanisms have been suggested for the initial formation of galaxies and clusters of galaxies: (1) Gravitational instability and (2) Turbulence in the early Universe.

1) Gravitational Instability

Let us consider that the whole Universe is filled uniformly with gas. A small local perturbation in the density of this gas may be enhanced or damped. Indeed, a local density excess causes a stronger local gravitational field which will tend to attract even more matter and thus will increase. On the other hand, the gas pressure will tend to disperse any density enhancement and restore the initial homogeneity. This problem was studied by *Jeans* in 1902. He found that small scale perturbations are quickly dispersed, while large scale perturbations become enhanced. In the second case, the density in the perturbation increases continuously with time. This is called "gravitational instability" or "Jeans' instability". This instability finally creates a concentration of matter which may evolve to form a star, a galaxy, or even a cluster of galaxies. The amount of matter condensed in this way depends upon the initial density of the gas and the local sound speed, at which speed density perturbations propagate. The minimum mass required for the onset of gravitational instability is called the "Jeans' mass", M_j, and its radius is known as the "Jeans' radius", λ_j.

In a sphere of radius greater than λ_j, gravity overcomes the gas pressure and causes a concentration of matter (Sect. 8.4). The reverse happens for a sphere with radius less than λ_j; the pressure of the gas overcomes gravity and the perturbation is damped.

Before the recombination era ($t < t_{rec}$) the Jeans' radius was very large because the sound speed at that time approached the speed of light, since matter and radiation were strongly coupled. The Jeans' mass increased until shortly before recombination, when it was approximately $10^{17}\ M_\odot$, that is, much greater than the mass of a cluster of galaxies.

After recombination ($t_{rec} = 700\,000$ years), matter and radiation were decoupled and the radiation ceased contributing to the pressure. Then the sound speed suddenly fell to a few kilometers per second. The corresponding Jeans' mass also dropped to $10^5\ M_\odot$, that is, comparable to the mass of a globular cluster.

We distinguish two extreme types of condensations of matter: (a) isothermal, and (b) adiabatic. In the former case, the temperature inside the perturbation is the same as the cosmic temperature. This is achieved by the free movement of photons which remain uniformly distributed, while the matter is clumped. In the latter case, the ratio of photons to baryons is the same inside and outside the perturbation, so that the temperature increases along with the density. That is, the ratio of radiation to matter does not change, but the temperature in the perturbation does. Each type of perturbation has its own implication for the future evolution. Consequently, two different theories have been proposed for the formation of galaxies and clusters of galaxies: the theory of the isothermal perturbations and the theory of the adiabatic perturbations.

a) Theory of the Isothermal Perturbations

This theory has mainly been proposed by *Peebles* (1965). It states that any isothermal perturbation in the initial distribution of matter in the Universe does not evolve before the time of recombination. The perturbations simply follow the expansion of the Universe. After t_{rec}, however, every perturbation which is greater than the Jeans' mass M_j starts to grow. As we saw earlier, the Jeans' mass after recombination is equal to $10^5\ M_\odot$, the mass of a globular cluster.

After the formation of these condensations we have two opposite effects proceeding in parallel. On the one hand, these clusters break up into small condensations which ultimately form stars. On the other hand, the same clusters concentrate in larger and larger groups which make up the galaxies, groups of galaxies, clusters and superclusters of galaxies. The larger the scale of concentration, the longer it takes to be formed. Galaxies were formed relatively fast, over a period of 300 000 000 years. During this time, the matter in the galaxies became completely mixed. On the other hand, the larger condensations of matter, such as the superclusters, have not yet fully collapsed, and have not yet taken their final form.

There are two basic arguments supporting this theory. The first is that a study of the observational data on the galaxy distribution, performed by *Peebles* and his collaborators, shows that there are no distinctive scales for groups of galaxies up to superclusters. That is, there are concentrations of galaxies on various scales, without particular preference for a certain scale. The second argument is based upon numerical calculations done with models of the expanding Universe (*Aarseth, Miller*, etc.). They found that points initially uniformly distributed (each point representing a galaxy) tend to segregate into groups which tend to increase in size as the Universe expands.

The initial perturbations are thought of as being due to perturbations in the initial distribution of matter in the Universe during the period of the formation of the atoms. These small perturbations increase with time, forming concentrations which are similar to the clusters and superclusters we observe today. By the current age of the Universe ($t = 1-2 \times 10^{10}$ years), the distribution of "points" is similar to the observed distribution of galaxies in the Universe (Fig. 3.4).

An interesting aspect of this theory is that large areas void of galaxies are formed in between the concentrations of galaxies. As time passes, the voids increase in size while the concentrations become more compact. So this picture seems to explain several characteristics of the observed Universe.

b) Theory of the Adiabatic Perturbations

This theory has been proposed by *Zeldovich* (1970) and his collaborators. A characteristic property of adiabatic perturbations is that small condensations are destroyed by viscosity[1] during the epoch before recombination. Only concentrations more massive than $10^{13} M_\odot$ can survive until t_{rec}, when they can collapse since they are then greater than the Jeans' mass.

According to this theory, the large scale structure (superclusters) formed first, while structure on smaller scales (clusters, groups of galaxies and galaxies) were formed later on by the fragmentation of these initial concentrations. This is exactly the inverse process to the one suggested by the theory of isothermal perturbations.

The superclusters, according to this theory, are not even approximately spherical, but are flat like "pancakes". For this reason, this theory is also known as the "pancake theory". These "pancakes" eventually fragment into clusters and groups of galaxies, which in turn fragment into galaxies. This phase lasts about 1 000 000 000 years and is followed by the collapse to form galaxies, with the formation of stars coming last of all.

This theory also has several attractive characteristics. For example, in a picture of the distribution of galaxies one can distinguish several elongated structures, consisting of galaxies, groups of galaxies and clusters, which are reminisent of *Zeldovich's* pancakes (Fig. 3.4). The

---

[1] Viscosity is the sticky property of matter, and is related to the forces between neighbouring atoms or molecules. Before t_{rec} the viscosity was mainly due to photons.

numerical experiments are consistent with this picture if we assume that the points represent particles, rather than galaxies.

It is premature to say which of these two theories best describes the formation of galaxies and groups, clusters and superclusters of galaxies. What is common in both is that the initial perturbations in the distribution of matter in the Universe, which led to the formation of galaxies, were very small and appeared during the first stages of the expansion of the Universe. They probably existed already at the Planck time ($t = 10^{-43}$ sec) and evolved into protogalaxies. Thus the isotropy and homogeneity of the early Universe, according to these theories, was almost exact[2].

2) Turbulent Motions in the Early Universe

This theory was first put forward by *von Weizsäcker* (1951), but its current version is due to *Ozernoi* (1968) and his colleagues.

According to this theory there was a "cosmic turbulence", characterised by random motions, right from the beginning of the Universe. The implication is that the Universe was not isotropic. Even so, its density was initially constant, that is, the Universe was initially homogeneous and only later were fluctuations formed, called "eddies", which led to the formation of galaxies.

The size of the eddies grew during the radiation era, while during the matter dominated era it diminished. It is estimated that the largest size attained by the eddies, which were later to form the galaxies, was of the order of $10^{12}\ M_\odot$, i.e. as large as the largest galaxies we observe today.

Before t_{rec} the velocities of the eddies were much smaller than the sound speed which, as we mentioned earlier, approached the speed of light. After t_{rec}, however, the sound speed suddenly dropped to a few kilometers per second and so the turbulent motion became supersonic. As a consequence, "shock waves" were formed like those formed by supersonic aircraft. These shock waves created concentrations of matter.

This theory was very popular a few years ago. Today, however, it is rather out of favour as it has come under serious criticism. The main argument against it is that the eddies which are the major element of this theory should dissipate into smaller eddies, as is usually observed in turbulent motion. Further, it is not clear that the shock waves after t_{rec} form concentrations of the right mass range (masses of galaxies) or much smaller masses. Finally, the anisotropy in the early Universe, postulated by this theory, is in contradiction to the almost complete isotropy observed in the microwave background radiation.

Even so, the three theories of galaxy formation mentioned here are still under discussion and no definite conclusion has been reached. There are

---

[2] Nevertheless, these perturbations are not so small as to be considered entirely random. It seems that the initial perturbations had a certain non-random structure. This problem is related to the possible "inflation" of the Universe (Sect. 8.8).

a few other theories for the formation of galaxies, but the above mentioned are the most important ones.

We come now to the next stage of galaxy formation, the collapse of the protogalaxies.

8.3.2 Theories of Galaxy Formation from the Collapse of Protogalaxies

The formation of protogalaxies was only the first stage in the formation of galaxies. Indeed, the protogalaxies were amorphous blobs of matter, much larger than the present galaxies, and the problem now is to explain how galaxies, as we see them today, formed out of these blobs. It is particularly important to explain the various morphological types of galaxies we mentioned in Chap. 2. That is, we would like to know why some galaxies turned out to be elliptical, some spiral and others irregular.

The simplest explanation for the formation of galaxies out of protogalaxies is based upon the collapse of the protogalaxies. That is, the matter in the protogalaxies (stars or gas) moves rapidly towards the centre, without any opposing forces. The "collapse time" is equal to the "free-fall" time for the matter in the galaxies. It is estimated that this time is approximately 300 000 000 years for an ordinary galaxy. This timescale could be larger if the protogalaxies were significantly larger in size.

There are two scenarios which describe the collapse of protogalaxies to form galaxies. According to the first, the protogalaxies are basically made up of gas, while according to the second, they consist mainly of stars.

a) Collapse of Gaseous Protogalaxies

This theory has been suggested by *Larson* (1969, 1976). He considered a protogalaxy made up of gas which collapses. The gas consists of clouds which collide inelastically, that is they lose energy during each collision. During these cloud collisions particularly dense concentrations of matter were formed, which subsequently evolved into stars. Thus we have continuous star formation during the collapse. The rate of star formation depends upon the initial density and the random motions inside the protogalaxy. The higher the density and the random motions, the more stars are formed. Galaxies which form stars rapidly, turn into ellipticals in the Hubble classification (Sect. 2.1). After the collapse of these galaxies very little gas is left, in agreement with observations.

Larson's theory explains the major characteristics of elliptical galaxies, which are their high central density and the drop of density in the outer parts being inversely proportional to the third power of the distance from the centre ($\varrho \propto r^{-3}$). It also explains the chemical gradient observed in these galaxies (stars with more heavy elements are found nearer the centre). During the first stages of the collapse, the first generation of stars was formed, in particular in the central regions. These stars essentially consisted of hydrogen and helium. The most massive of them evolved faster, synthesising heavier elements (elements other than hydrogen and helium) in nuclear reactions. These stars eventually exploded and their heavy elements were spread in space, and made up the material from which younger stars, richer in metals, were formed.

Protogalaxies which were not very dense collapsed more slowly. Part of the gas did not have enough time to form stars during the collapse stage and accumulated in the plane perpendicular to the rotation axis of the galaxy, the plane of symmetry. In this way, a relatively thin layer of gas was formed, called the "disc" of the galaxy. Stars in the disc were formed much later and at a slower rate. These are the population I stars which are relatively young. On the other hand, stars which formed during the collapse are the oldest stars of the galaxy and are called population II stars.

This is the way that spiral galaxies were formed. We may say that spiral galaxies consist of two stellar populations, population II stars which form a structure like an elliptical galaxy, and population I stars which form a flat disc. Elliptical galaxies consist almost exclusively of population II stars. The same holds for the globular clusters. Finally there are the irregular galaxies, some of which mainly comprise population I stars while others mainly comprise population II stars.

The age of the globular clusters and elliptical galaxies (and, in general, all population II stars) is more than 10 billion years, while population I stars are much younger, usually 5–6 billion years or less.

The formation of population I stars takes place mainly in the spiral arms of the galaxies. The higher density of the matter in this region creates shock waves which increase the density even more. A consequence of this increase in density is the formation of new stars, either individually, or in associations and clusters (open clusters, which are much smaller and younger than globular clusters).

Spiral galaxies are apparently surrounded by an approximately spherical structure of large dimensions, called a "halo" (Fig. 1.2). The halo consists of faint population II stars and gas (and possibly black holes, neutrinos, or other exotic particles).

The halo plays an important role in the stability of galaxies. Galaxies without a halo tend to become elongated. There is evidence that elliptical galaxies are in general triaxial, that is, they lack any symmetry axis. Normal spiral galaxies (Figs. 1.2, 1.3 and 2.4, 2.5) seem to have a flat disc in which the spiral arms are formed. On the other hand, barred spirals have bars which are more elongated than elliptical galaxies (Fig. 2.6). It seems that the difference between the various types of galaxies is mainly due to the existence, or otherwise, of a halo. The halo, which is significant in the case of normal spiral galaxies, stabilizes the disc and does not allow the formation of a bar.

Recapitulating, we explain the formation of the various types of galaxies in the Hubble classification (Sect. 2.1) as follows. Elliptical galaxies exhausted their initial gas supply during the collapse, and essentially consist only of stars. Spiral galaxies have a gaseous disc which is left over from the initial collapse. The normal spiral galaxies have a large halo which stabilizes the disc, while it seems that in barred spirals the halo is smaller. Finally, the irregular galaxies are usually small and they do not seem to have followed a regular evolution as the larger galaxies.

b) Collapse of Stellar Protogalaxies

This theory is different from *Larson's* theory, since it assumes that stars, which make up the elliptical galaxies, were formed before the collapse of the protogalaxies. It has been developed by *Gott* (1973–1975) and is mainly based on numerical computations of *N*-body collapses. The expansion of the Universe is also taken into account in these computations. *Gott* noticed that after the first collapse, the outer layers of the protogalaxy, which might still be expanding, were decelerated by the partly formed galaxy, and turned around and collapsed as well, thus producing the extended outer envelopes of these galaxies, where the density drops like the inverse radius cubed.

This theory, however, does not explain the high central densities of elliptical galaxies. Further, it cannot explain the formation of spiral galaxies because, unlike gas, stars cannot form a thin disc.

Even so, it is not out of the question that some elliptical galaxies might be better understood if one accepts that their stars were formed before the collapse of the protogalaxy. As we mentioned earlier, the possibility that there are stars which are even older than population II has recently been considered seriously. These are the population III stars which are believed to have been formed even before the protogalaxies.

c) Other Theories

Another way of forming certain types of galaxies is by collisions of galaxies (*Toomre* 1977). For example, if two roughly spherical galaxies collide, they may form a prolate elliptical galaxy. Also, it is possible for a large galaxy to absorb smaller galaxies it encounters. This is called "galactic cannibalism" and provides an explanation for some giant galaxies observed in the centres of clusters of galaxies. These galaxies most likely have absorbed many small galaxies which have collided with them.

It has been suggested that quasars, which show signs of intense activity (explosions) in their nuclei, might be galaxies in formation. If this is so, all galaxies must have passed through the quasar stage.

The current research is aimed in two directions. Firstly towards a better theoretical understanding of the evolution of galaxies. The theories we mentioned earlier give us a basic understanding of galaxy formation, and the problems which are currently being studied try to answer particular questions. Secondly, much effort has gone into trying to discover relatively young galaxies. It is estimated that some galaxies, among which are, most likely, the quasars, have not yet reached their final state of equilibrium. Therefore, studying such galaxies may enable us to put constraints upon the various theories of galaxy formation.

8.4 Jeans' Instability

We are going to derive here the "Jeans' length", which characterises the "Jeans' instability" or "gravitational instability". The gas pressure tends to disperse any concentration of matter while gravity tends to enhance it. Pressure disperses the gas at the sound speed v_s, while gravity enhances the gas density at the collapse velocity v_c. If v_c is greater than v_s, gravity dominates and gravitational instability results, that is, the initially small perturbation grows.

The potential energy of a particle on the surface of a sphere with radius r, is equal to

$$V = \frac{GMm}{r} = \frac{4}{3}\pi Gm\varrho r^2,$$ (8.7)

where M is the mass and ϱ the density of the sphere, m is the mass of the particle and G the gravitational constant. This potential energy must be equal to the kinetic energy of the particle at the centre of the sphere

$$T = \frac{mv_c^2}{2}.$$ (8.8)

Therefore the velocity of the collapse is

$$v_c = \sqrt{\frac{8\pi G\varrho}{3}}\, r.$$ (8.9)

If v_c is equal to the sound speed v_s, r will be equal to the "Jeans' length"

$$r = \lambda_j = \sqrt{\frac{3}{8\pi G\varrho}}\, v_s.$$ (8.10)

More accurately, the "Jeans' length" is defined as:

$$\lambda_j \equiv \sqrt{\frac{\pi}{G\varrho}}\, v_s$$ (8.11)

which is different from (8.10) only by a numerical factor.

8.5 Cosmology and Elementary Particles

A lot of effort has been made, in recent years, to unify the two extreme areas of research, cosmology, which deals with large scale phenomena in the Universe, and microphysics, which studies the minutest constituents of nature, in particular the elementary particles.

The effort to unify the various theories of physics has always been one of the major motivations of scientific research. The most important achievements of science have been exactly those theories which managed to unify branches of physics which at first appeared to be entirely unrelated. Thus, *Newton*'s theory in 1666 unified the fall of particles to the Earth with the movement of the celestial bodies. Then we had *Maxwell*'s theory of 1873 which unified magnetism and electricity, and showed that light, as well as the invisible radiations, infrared, ultraviolet, x- and γ-rays, are nothing but electromagnetic waves. Atomic physics followed (*Bohr, Schrödinger, Heisenberg* and others, from 1913 onwards) which showed that atoms and molecules obey the laws of quantum physics, and their behaviour is governed by electromagnetic forces. Thus all the chemical and biological phenomena could be explained using the basic laws of atomic physics, based on the electromagnetic forces.

Another big step in the unification of physics was made by *Einstein*'s theory of relativity (1905 special theory, 1915 general theory). This theory unified space and time into one entity, spacetime, and showed that gravity is nothing other than an expression of the curvature of this spacetime. This theory also brought about the unification of energy with matter through the formula $E = mc^2$.

Einstein's success led him to attempt to unifiy gravity with electromagnetism. So, during the last 20 years of his life, he tried to describe electromagnetism as a property of spacetime just like gravity. The result was his Generalised Theory of Gravitation (1950) which, however, as is generally known, was not successful.

We come now to the present time, with the theories for the unification of the electromagnetic and the "weak" nuclear forces, by *Weinberg, Salam* and *Glashow* (Nobel prizes 1979) as well as the "grand unified theories" of *Weinberg, Glashow, Georgi, Nanopoulos* and others. These theories also attempt to unifiy the "strong" force of microphysics and possibly gravity as well. Their major characteristic is that they start from microphysics in contrast to *Einstein* who started from gravity, which governs the whole Universe.

It is now well known that there are four basic forces in nature:

1) Strong (Nuclear) Force. The original definition of the strong force refers to the attraction between nucleons (protons and neutrons). These particles are attracted strongly when their distance is about 10^{-13} cm, i.e.

equal to the dimensions of a proton. The strong force binds the nucleus together. At larger distances, the protons repel each other by electric forces. But if we can bring the protons very close to each other the attractive force is much stronger and sticks the particles together.

At present, the term "strong force" is used to denote the attraction between quarks inside the hadrons, like nucleons and mesons. Therefore, the strong force is the force that keeps the quarks of a proton, or of a neutron etc., together. The attraction between two neighbouring hadrons is much weaker and is due to the strong force acting between the quarks in the two particles.

2) Weak (Nuclear) Force. The weak force dominates at distances of 10^{-16} cm and is mainly responsible for the decay of the neutron. Indeed, when neutrons are free, they are unstable, decaying into a proton, an electron and a neutrino, with a half life of 15 minutes. The electrons produced by such decays constitute the β-radiation from the radioactive particles.

3) Electromagnetic Force. This force operates between charged particles (protons, electrons, etc.). At small distances it is negligible in comparison with the strong nuclear forces, but at distances greater than 10^{-13} cm it is much stronger. It is responsible for the formation of atoms and molecules, thereby being responsible for the structure of matter on the atomic and laboratory scales. Organic and inorganic chemistry, and in particular life itself, are governed by this force.

4) Gravitational Force. The gravitational force is much weaker than the electromagnetic force between two particles (for a proton and an electron it is 10^{40} times weaker). Therefore, gravitational forces between atoms are entirely negligible in comparison with the electromagnetic forces. When, however, the masses of the various bodies are large, the gravitational forces are enormous. This is because the matter in the celestial bodies (planets, stars and galaxies) is neutral, that is, the celestial bodies are not charged, or at least their charges are exceedingly small. Therefore, the electromagnetic force between two stars is negligible, while the gravitational force dominates. Thus in the case of large masses, gravitational attraction dominates over all other forces. For example, if a star has a mass which is greater than 3 solar masses, it will collapse to form a black hole, and no other forces (electromagnetic, weak or strong) can stop this collapse.

The gravitational and electromagnetic forces are "long range" forces, that is, they can be felt at large distances, in contrast to the first two forces which are only felt at distances less than 10^{-13} cm, i.e. at distances equivalent to the dimensions of protons and similar elementary particles.

Particles which interact via the strong force are called "hadrons". Such particles include the protons and neutrons that constitute the nuclei (which is why they are also called nucleons), the hyperons (particles which are heavier than protons) and the mesons (particles like the π meson, which are lighter than the proton). Other particles, however, like the electrons (e) and neutrinos (v) do not interact via the strong force. Such particles are called leptons. Further, the neutrinos do not even interact via the electromagnetic force.

Photons are distinguishable from all other particles, mainly because of the special role they play in the transmission of the electromagnetic interactions. As is known, the electromagnetic effects travel in the form of waves and according to quantum mechanics, every type of wave corresponds to a particle. The quantum of the electromagnetic radiation is the photon, which has zero rest mass and moves with the speed of light.

Similarly, the quantum of gravity is called the graviton. As we shall see later, both the weak and the strong interactions have their corresponding quanta (interaction particles).

The various elementary particles are characterised by their rest mass, and certain quantum numbers. Such quantum numbers are the charge Q, the spin [3], the baryon number B and the lepton number E. These numbers remain constant (exactly, or approximately) during the various nuclear reactions.

Every particle has a corresponding antiparticle. For example, the proton with charge $+1$ and baryon number $+1$ corresponds to the antiproton, with charge -1 and baryon number -1; the electron corresponds to the positron, the neutrino to the antineutrino, etc.

Apart from the elementary particles we mentioned above, there are plenty more which have been discovered by the large accelerators. Today, hundreds of such particles are known, most of which are entirely unstable, with half-lifes which are a small fraction of a second. These particles decay to simpler ones, like the nucleons, the mesons or the leptons.

Therefore, there are many more elementary particles than was originally thought and their classification is a great problem. The one hundred or so known chemical elements can be explained by means of their basic constituents, protons, neutrons and electrons. However, if we consider all the elementary particles that can appear in the atomic nuclei, we have a real "chaos".

The quark theory manages to put some order into this "nuclear chaos". The theory was proposed in 1961 by *Gell-Mann* (1961; Nobel prize 1969), and independently by *Neeman* (1961).

According to the quark theory, all hadrons consist of even more elementary particles, the quarks. These particles have spin 1/2 and electric

---

[3] The elementary particles have either integer spin, in which case they are called bosons, or half integer spin, in which case they are called fermions.

Table 8.2. Quarks

Quark	Spin	Charge	Baryon number
u	1/2	2/3	1/3
d	1/2	$-1/3$	1/3
s	1/2	$-1/3$	1/3
c	1/2	2/3	1/3
t	1/2	2/3	1/3
b	1/2	$-1/3$	1/3

charge 1/3 or 2/3 of the elementary electronic charge. *Gell-Mann*'s theory postulated 3 quarks, the u (for up), d (for down) and s (for strange). Table 8.2 gives their major characteristics. Further, there must be an equal number of antiquarks which characterise the antimatter, the ū, d̄ and s̄. The proton consists of two u quarks and one d quark, while the neutron consists of two d quarks and one u quark. A meson π^0 consists of one u quark and one antiquark, ū. In general, the baryons consist of three quarks, while the mesons of one quark and one antiquark. All elementary particles are made up of similar combinations of quarks so that their charge is an integral multiple of the electronic charge.

Further, quarks exist in three different forms, called "colours". So we have "blue", "green" and "red" quarks. The combination of the quarks which make up a baryon must be such that the three colours should be in equal quantities, and their combination "blue + green + red" is equivalent to "white". That is, baryons do not have "colour". Of course, the word "colour" has nothing to do with the optical colours. It simply expresses a property of the quarks.

With the postulation of the quarks, a new quantum number was added to those we have already mentioned, the "strangeness" *S*. Particles with $S \neq 0$ are called strange. Later on, theoretical considerations by *Glashow, Maiani* and *Iliopoulos* (1970) showed the existence of a fourth quark, called c (for charmed), which is characterised by a new property, the "charm", and has a corresponding quantum number. The existence of the c quark was proven in 1974 with the discovery of the ψ particle (or *J* as it is otherwise known; sometimes also referred to as J/ψ) which consists of one c quark and its antiquark. Because of this discovery *Ting* and *Richter* were awarded the Nobel prize in 1976.

Recently, two more quarks have been introduced, the t (for top) and b (for bottom). The discovery, in 1977, of the υ particle, which consists of one b quark and its antiquark, confirmed the existence of the b quark. Finally, experiments at CERN during 1984–1985 indicate the existence of the t quark. It is thought that at most two more quarks might exist.

Many efforts have been made so far, to isolate the quarks. In spite of all this effort, no quark has yet been observed in isolation. It seems that the forces which hold them together increase as the quarks move away from each other. It appears, therefore, impossible to isolate one quark; quarks can only exist in bound states, inside hadrons.

Every pair of quarks corresponds to a lepton and a corresponding neutrino. Thus, the quark pair (u, d) corresponds to the electron (e^-) and the electron neutrino (v_e); the quark pair (s, c) corresponds to the muon μ[4] and the muon neutrino (v_μ); and the quark pair (t, b) corresponds to the heavy lepton (τ) and its corresponding neutrino (v_τ).

The maximum number of quarks can be established from cosmological constraints. During the first four minutes, when the first elements of matter were created, the amount of helium formed depended upon the ratio of neutrons to protons n/p. Indeed, all neutrons reacted with protons and formed deuterium D, which then formed $_2He^4$ (Fig. 6.5). Therefore, the higher the ratio n/p the more helium was finally formed. The ratio n/p depended upon the density of matter ϱ when the nucleosynthesis started, i.e. when the temperature was approximately 10^9 K. The more types of particles existed in the Universe, the higher the density ϱ would have been. More specifically, every type of neutrino contributed equally to ϱ. Therefore, the more neutrinos we had, the more $_2He^4$ would have been formed.

The best estimate we have for the amount of helium $_2He^4$, formed during the first stages of the expanding Universe, is about 25% of all the mass in the Universe. This figure implies that there are most likely only three neutrinos in the Universe, which in turn implies that there are only three pairs of quarks in total. One may expect, at most, one more pair of quarks if the uncertainty in our calculations is taken into account.

Let us go back now to the various efforts to unify the forces of nature. The greatest success in recent years has been the theory by *Weinberg* (1967) and *Salam* (1968), which improved upon an older theory by *Glashow* (1961). This theory unifies the electromagnetic and the weak forces. The unified force is now called "electroweak". There is already enough experimental evidence to support this theory. Its major prediction was that the weak forces are transmitted by some special quanta called "intermediate bosons". These quanta may be charged (W^+, W^-) or neutral (Z^0) and they are very massive, their masses being of the order of 100 GeV/c^2 (that is, masses which correspond to energies of 100 billion electron volts). Such masses are 100 times greater than the mass of the proton. More accurately, the mass of W^+ and W^- is approximately 82 GeV/c^2 while the mass of Z^0 is approximately 93 GeV/c^2. These particles were indeed discovered at the beginning of 1983 in CERN during experiments of proton-antiproton collisions. Their masses were found to be approximately the theoretically

---

[4] The muon is also called a μ meson, for historical reasons, since it was the first particle to be discovered with a mass between that of the proton and the electron. Nowadays, however, the name mesons is reserved for the hadrons which consist of one quark and one antiquark, while the μ particle is a lepton, that is a point charge, and not a composite particle like the π meson.

predicted ones. This verification is one of the greatest discoveries of recent years. The 1984 Nobel prize was awarded to *Rubbia* and *van der Meer* for this discovery.

The electromagnetic and weak forces in the Universe were indistinguishable when the age of the Universe was less than 10^{-12} sec (electroweak time, Fig. 8.1), and its temperature greater than 10^{16} K.

The strong interactions, on the other hand, occur via certain particles, the quanta of the strong field, which hold the quarks together. These quanta are called gluons. Gluons exist inside the elementary particles and, just like quarks, cannot be seen outside them. It is considered that two quarks are attached together with the constant exchange of gluons. There are 8 types of gluons. Six of them change the colour of the quark (from blue to green or red, from green to blue or red and from red to blue or green), while the remaining two gluons leave the colour unchanged.

The theory which deals with these interactions between quarks, mediated by the gluons, is called "quantum chromodynamics", QCD, in analogy with "quantum electrodynamics", QED, which has so far dealt very successfully with the electromagnetic interactions. Quantum electrodynamics assumes that the force between two charged particles is due to the exchange of a photon between them. By analogy, in quantum chromodynamics the role of the charge is played by the colour, and the role of the photon by the gluon. Quantum chromodynamics today is one of the most active fields of research. It is much more complicated than quantum electrodynamics, first because it deals with much stronger forces, and second because gluons, as opposed to photons, interact with each other.

The recent efforts to unify the forces of nature are embodied in the "Grand Unified Theories" (GUT, *Weinberg, Glashow, Georgi, Nanopoulos,* etc.). These theories attempt to unify the "electroweak" force with the strong force. According to them, quarks and leptons belong to the same "large groups". In this way we can explain the correspondence between quarks and leptons.

A first consequence of this correspondence is the quantization of the electric charge. More specifically, such a correspondence explains why the charge of the proton is exactly the same as the charge of the positron.

A second consequence of this correspondence is the ability of quarks to form leptons. This can happen when the distances between the quarks reduces to below 10^{-29} cm (compare this with the radius of a proton which is 10^{-13} cm). Such a reaction, which violates the conservation of baryon number B, may occur in a "grand unified field" with the exchange of an extra heavy particle X (Fig. 8.2). The mass of this particle is approximately 10^{15} GeV/c^2, that is 10^{15} times the proton mass. Such heavy particles cannot be observed experimentally in our accelerators. Their existence may be checked, however, indirectly. The major consequence of the existence of these particles of the "grand unified field" is that protons (and all hadrons) are unstable.

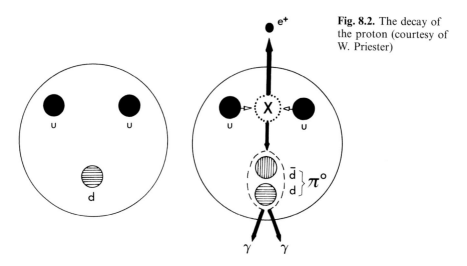

Fig. 8.2. The decay of the proton (courtesy of W. Priester)

The probability that two quarks pass within 10^{-29} cm and thus trigger the decay of the proton is very small. It is estimated that the half life of the proton is enormous, of the order of 10^{31} years. Today, experimentalists have made a lot of effort to observe the decay of the proton[5]. If this decay is confirmed, the implication is that one day all hadrons in the Universe will decay.

Protons may decay by either of the two reactions:

$$p \rightarrow \pi^0 + e^+, \quad \text{or} \tag{8.12}$$

$$p \rightarrow \pi^+ + \bar{\nu}. \tag{8.13}$$

The neutral pion π^0 decays into two photons,

$$\pi^0 \rightarrow \gamma + \gamma, \tag{8.14}$$

while the positive pion π^+ decays into a positive muon and a neutrino:

$$\pi^+ \rightarrow \mu^+ + \nu. \tag{8.15}$$

The positive muon, μ^+, further decays into a positron, a neutrino and an antineutrino:

$$\mu^+ \rightarrow e^+ + \nu + \bar{\nu}. \tag{8.16}$$

---

[5] After some preliminary claims that proton decay had been observed, more recent experiments (1983) show that the proton's half life is at least 10^{32} years.

The positron finally may be annihilated with an electron to produce two photons:

$$e^+ + e^- \rightarrow \gamma + \gamma. \tag{8.17}$$

On the other hand, the neutrons are known to decay to a proton, an electron and an antineutrino:

$$n \rightarrow p + e^- + \bar{\nu}. \tag{8.18}$$

In a similar way all other hadrons may decay too, so that finally all the matter of the Universe will be converted into photons and neutrinos.

Another important application of the "grand unified theories" to cosmology is the explanation of why the Universe consists mainly of matter and not antimatter, or a mixture of both.

When the age of the Universe was less than 10^{-35} sec (GUT time, Fig. 8.1), matter and antimatter existed in roughly equal proportions. During that period, the temperature was greater than 10^{27} K and the density greater than 10^{80} g/cm^3. Under such conditions processes which violate the conservation of baryon number were possible. These processes are thought to have been mediated by the X particle, and its antiparticle $\bar{X}$. At temperatures greater than 10^{27} K these particles existed in roughly equal numbers, being in thermal equilibrium with the other particles. As the temperature dropped below 10^{27} K, the X and $\bar{X}$ particles decayed to produce baryons and antibaryons, respectively. If, however, the CP symmetry was violated at the same time (Sect. 9.1.4) the decay rates of X and $\bar{X}$ would have been different, resulting in a small excess of baryons over antibaryons, i.e. matter over antimatter. Since, then, the change in the baryon number has been very slow, and is now due solely to the proton decay.

After the baryon number had become different from zero, the reactions between matter and antimatter that followed eliminated practically all the antimatter, and left only a certain quantity of matter. The residual matter was much less than the matter which had been annihilated, but it was enough to form the galaxies and stars later on.

The subject of antimatter in the Universe is particularly interesting and it will be examined in detail in the following section.

8.6 Matter and Antimatter

How much antimatter exists in the Universe? This is a particularly interesting question because it concerns the initial stages of the Universe.

During the first one and a half millionths of a second, the temperature was so high that photons could create pairs of nucleons and antinucleons (that is, protons and antiprotons, neutrons and antineutrons, etc.). The formation of electrons and positrons (antielectrons) lasted much longer, for almost 5 seconds, because their mass is so much smaller, and therefore the photons required to produce them could be much less energetic. So, during the first stages of the Universe, the amount of antimatter was, if not equal, at least approximately equal to the amount of matter. When the temperature dropped enough for the production of pairs of particles and antiparticles to cease, the various particles were continuously being annihilated by collisions with their antiparticles, producing γ-rays. Thus, the number of particles and antiparticles decreased rapidly, with a corresponding increase in the number of photons.

We can estimate theoretically how many particles and antiparticles existed at any time. If we initially had equal numbers of nucleons and antinucleons, we should have 10^{-19} nucleons and antinucleons per photon today.

What we observe in our Galaxy, however, is very different. There are 10^{-9} nucleons per photon, and an insignificant number of antinucleons.

There are two possible explanations for this significant discrepancy between theory and observations: Either there were many more nucleons than antinucleons to begin with, or in some way the nucleons and antinucleons became separated in the early stages of the expansion, so that they avoided annihilating each other. If the latter is true, there must be, somewhere in the Universe, an equal number of celestial bodies made from antimatter as from matter.

The theory of a symmetric Universe, with respect to matter and antimatter, has been proposed by *Klein* (1958) and *Alfven* (1965), and later on, with more details, by *Omnes* (1972) and his collaborators. The theory by *Klein* and *Alfven* is radically different from the Big Bang theory. It assumes that the Universe was initially diffuse and *contracting* due to gravity. When matter and antimatter came together, annihilation started on a large scale. The pressure created in this way stopped the contraction and caused the well known expansion. This theory does not have much appeal because it does not explain the microwave background radiation, and the other characteristics of the Universe. *Omnes'* theory, on the other hand, is based upon the Big Bang theory. It accepts that matter and antimatter became separated during the early stages of the expansion. Galaxies and stars were formed in each of these separated regions, so that there is today an equal number of galaxies made up of matter as of antimatter.

The matter-antimatter theory has a theoretical argument in support of it: Almost nothing in nature shows a preference for matter over antimatter. Also, the principle of the conservation of baryon number (the baryon number is equal to the number of baryons minus the number of antibaryons) shows that if it were zero in the initial singularity, it should still be zero today. Therefore, the amounts of matter and antimatter in the Universe today must be equal.

The properties of the atoms of antimatter are no different from the properties of the atoms of matter. Just as the atoms of matter consist of nuclei made from protons and neutrons with electrons around them, atoms of antimatter consist of nuclei made from antiprotons and antineutrons with positrons around them.

The energy levels of the positrons in the antiatoms are the same as the energy levels of the electrons in the atoms. Therefore, their spectral lines are the same. So, spectroscopic observations cannot help to distinguish between objects made from matter or antimatter. The same is true for all gravitational and electromagnetic phenomena. Therefore, a system made from antimatter behaves just like a system made from matter. Antimatter can only be observed when it reacts with matter.

Let us now examine what the observations tell us about the existence of antimatter in the Universe, according to *Steigman* (1976).

It is obvious that the amount of antimatter on the Earth is negligible. Of course, antiparticles are produced during nuclear reactions, or during collisions of the cosmic ray particles with those of the atmosphere, but these antiparticles are quickly annihilated in collisions with particles. Only with the help of special techniques, like the use of strong magnetic fields, can we maintain antiparticles for long periods of time. Thus, we can generate rings of rotating positrons or antiprotons which are used for various nuclear reactions. In any case, the amount of antimatter on the Earth is entirely insignificant in comparison with the amount of matter.

The same holds for the rest of the solar system. Our spaceships, which have landed on the Moon, Mars and Venus would have been instantly annihilated with a big explosion, if these celestial bodies were made of antimatter. The Sun is also made of matter and not antimatter. Indeed, solar material (the solar wind) continuously bombards the Earth. If it were made of antimatter, a lot of γ-rays would have been produced, contrary to observations. The solar wind also hits all the other planets, from which we ought to receive γ-rays if the planets were made of antimatter. This is again contrary to the observations. There is, therefore, no doubt that our solar system is made up of matter and not antimatter.

There are similar arguments which prove that the whole of our Galaxy is made from matter and not antimatter:

a) If there were stars made of antimatter in our Galaxy, there would also exist interstellar antimatter. Because of the motion of the interstellar material, we would then have matter-antimatter reactions, which would be observed as strong γ-ray sources. Of course, we do receive γ-rays from space, but this radiation is much less than there would be if there were appreciable amounts of antimatter present. Besides, the γ-rays we do receive can be explained in other ways, mainly as the products of various nuclear reactions. The conclusion is that the possible annihilation of matter and antimatter in our Galaxy is minimal. Thus we derive a maximum possible ratio for antimatter to matter of 10^{-15}. So, the amount of antimatter in our Galaxy is insignificant.

b) We receive cosmic rays from stars of various types, mainly novae, supernovae and pulsars. If these were stars made from antimatter, we would receive primordial cosmic ray radiation made from antimatter, contrary to what we observe. We only receive a few antiprotons from nuclear reactions that occur along the route of the cosmic rays. If these antiparticles were coming from sources of antimatter, we would also expect to see a certain quantity of α-antiparticles (that is nuclei of antihelium), or even of heavier elements of antimatter, in the same proportion that we find them in cosmic rays of matter. Until now, no such heavy antiparticles have been observed. This negative conclusion offers us an upper limit to the amount of antimatter in the Galaxy of 10^{-4} (ratio of particles of antimatter to those of matter).

We can now proceed to other galaxies. We notice that in clusters of galaxies there are collisions of galaxies, but the main interactions are observed in the intergalactic medium which emits x-rays. If these reactions were due to matter-antimatter collisions, we would observe γ-rays as well, which are not observed. It is estimated that the amount of antimatter in clusters of galaxies is less than 10^{-6} that of the matter.

These observations suggest that only on the scale of clusters of galaxies can we have separation of matter and antimatter. That is, current observations are consistent with the existence of clusters made from antimatter, in parallel with clusters made from matter.

There are, however, theortical difficulties in separating clusters of matter and antimatter in the early Universe. No theory has so far explained how such large regions of matter and antimatter became separated out of a fluid which contained equal numbers of particles and antiparticles. This is the major disadvantage of theories which accept equal amounts of matter and antimatter in the Universe. For this reason, it is now currently accepted that in the first fraction of a second in the expansion of the Universe (about 10^{-35} sec) more matter was formed than antimatter.

To explain this, we must appeal to the violation of the principle of the conservation of baryon number. We know at least one case where the baryon number is not conserved, namely black holes. Anything falling into a black hole does not leave any trace behind from which we can tell its baryon number B. This is a corollary of the "no hair theorem" (Sect. 5.10). According to this theorem, black holes can only be characterised by their mass, angular momentum and their charge. That is, a black hole will be the same when particles of matter (positive B) or particles of antimatter (negative B) fall into it.

The baryon number was also not conserved in the early Universe, when its age was about 10^{-35} sec. This stage of the Universe can only be described by the most recent cosmological theories, in particular, the grand unified theories which unify the strong nuclear forces with the weak and electromagnetic forces.

These theories predict the conversion of baryons into leptons, and therefore the violation of the conservation of both the baryon and lepton numbers. This results in a lack of symmetry between matter and anti-matter in the Universe. In contrast to the baryon and lepton numbers, the charge and mass are conserved quantities both in black holes and in the Universe at large. This difference in behaviour of the charge and mass on the one hand, and the lepton and baryon numbers on the other, is explained in terms of the forces which are involved: The mass and the charge are related to long-range forces (gravitational and electromagnetic), while the baryon and lepton numbers are related to the very short range forces (the strong and the weak nuclear forces).

The major problem, which is still unresolved, is the observed proportion of matter to light, i.e. why there are 10^{-9} nucleons per photon in the Universe. One of the first scientists who attempted to explain this number in terms of the violation of *CP* symmetry (Sect. 9.1.4) was *A. Sakharov* (1967). This question, however, remains open, and the observed matter/photon ratio has not been fully understood yet. So, there is no doubt that the current theories have not yet provided all the answers to the basic questions concerning the Universe. We are living through a period of intense activity in the field which combines cosmology with microphysics.

8.7 Quantum Gravity

A lot of discussion has been going on, in recent years, on the possible modifications of the general theory of relativity, during the first instant of the Universe, when its age was less than the Planck time, $t_p = 10^{-43}$ sec.

The Planck time is defined as follows: *Heisenberg*'s principle of uncertainty tells us that the product of the uncertainties in two conjugate quantities, for example the co-ordinate x and its momentum $\dot{x}$, is equal to Planck's constant $\hbar = h/2\pi$.[6] Two such conjugate quantities are the energy E and the time t. So, the principle of uncertainty gives

$$\Delta E \cdot \Delta t \simeq \hbar. \tag{8.19}$$

---

[6] Usually we call "Planck's constant" the elementary action $h = 6.625 \times 10^{-27}$ erg · sec. Quite often, however, the quantity $\hbar = h/2\pi$ is used and is also called Planck's constant.

If the uncertainty in the energy is equal to

$$\Delta E = 2mc^2, \tag{8.20}$$

where c is the speed of light, then in time Δt defined by (8.18) we may have creation and annihilation of two particles each of mass m. This time is of the order of

$$\Delta t = t_c = \frac{\hbar}{4\pi mc^2}, \tag{8.21}$$

and is called the "Compton time".

When the mass m is very large, the time t_c is very small, usually less than the "Schwarzschild time" which is defined as the time required for light to cross the Schwarzschild radius

$$t_s = \frac{2Gm}{c^3}. \tag{8.22}$$

For example, if m is equal to one solar mass, $t_c = 3 \times 10^{-82}$ sec and $t_s = 10^{-50}$ sec. But for small masses, the Compton time is greater than the Schwarzschild time. The two times are equal when the mass m has the value:

$$m_p = \sqrt{\frac{\hbar c}{8\pi G}} = 10^{-5} \, \text{g}. \tag{8.23}$$

This is the "Planck mass" and the corresponding time is the "Planck time"

$$t_p = \sqrt{\frac{G\hbar}{2\pi c^5}} = 10^{-43} \, \text{sec}, \tag{8.24}$$

while the corresponding length ("Planck length") is

$$R_p = \sqrt{\frac{G\hbar}{2\pi c^3}} = 10^{-33} \, \text{cm}. \tag{8.25}$$

Obviously, Planck's time is relevant to quantum effects on the one hand (*Heisenberg*'s uncertainty), and to gravity on the other (black holes). It can be shown that a black hole with a mass equal to the Planck mass cannot live for longer than the Planck time t_p because of the Hawking effect (Sect. 5.11). Therefore, black holes with masses less than or equal to the Planck mass radiate away all their matter, and disappear in a time which is less than or equal to the Planck time.

These properties of the Planck time imply that when the age of the Universe was less than $t_p = 10^{-43}$ sec, its behaviour was determined by quantum effects. So, many people tried to develop a quantum theory of gravity, which would replace the general theory of relativity during the very early Universe, i.e. when its age was less than 10^{-43} sec.

At the Planck time the density of the Universe was 5×10^{93} g $\cdot$ cm^{-3}, its temperature 10^{32} K and the average energy of its particles 10^{19} GeV. In general, the properties of matter then were entirely different from what they are now.

One of the major quantum phenomena during the "Planck era" was the creation of pairs of particles. This phenomenon has been studied by *Parker* (1969) and, in more detail, by *Zeldovich* (1972) and his collaborators. The creation of pairs of particles was due to irregularities in the gravitational field during the Planck time.

It is obvious that every elementary particle creates around it a strong gravitational field. The reverse, however, is also true: a strong irregularity in the gravitational field may create an elementary particle. This is a quantum phenomenon. Only the known elementary particles can be created (not particles with just any mass), and only in pairs (particle + antiparticle). Because of the creation of pairs of particles the irregularities in the early Universe were quickly smoothed out. That is, if the Universe were initially very irregular, it very quickly became homogeneous and isotropic, thanks to the creation of pairs of particles. This is considered to be one of the main mechanisms that caused the increase of the entropy to the rather high value of 10^9, in the early Universe. Thus the Universe already became, to a high degree, homogeneous and isotropic by the Planck time, i.e. when its age was only 10^{-43} sec. The study of the properties of the Universe during the Planck era, however, is still in its infancy. We have not yet managed to come up with a theory of quantum gravity which explains, in a satisfactory way, this era.

Another mechanism which could smooth out the irregularities in the very early Universe is neutrino viscosity. During the first stages in the life of the Universe, the neutrino density was enormous, and the neutrinos were behaving like a thick fluid with viscosity. Thus, the density and the temperature became uniform everywhere, so that the Universe quickly became homogeneous and isotropic.

The disappearance of irregularities in the Universe made its entropy increase. The entropy created during the Planck time was very high (10^9 photons per baryon). From $t = 10^{-43}$ sec till today ($t = 1 - 2 \times 10^{10}$ years) the entropy has been increasing continuously, but much more slowly, and its value has not risen substantially above 10^9 photons per baryon.

Theoretical calculations show that if the Universe were appreciably inhomogeneous before the Planck time, the creation of particles and the neutrino viscosity would have caused a much higher increase in the entropy than that observed. Therefore, the Universe was only slightly inhomogeneous during the Planck era.

Another theory relevant to Planck's era is the one concerning "primordial black holes", proposed by *Hawking*. These black holes are supposed to have been formed by the local collapse of the matter in points where the density of the Universe was very high, or the gravitational field very strong. Their masses were very low, from 10^{-5} g to 10^{15} g, i.e. much less than the mass of the ordinary black holes formed by collapsing stars, which have masses at least 3 times the solar mass. Primordial black holes may radiate and finally explode (Sect. 5.11). However no such explosion have been observed up to now.

In recent years much effort has been made to try to unify gravity with the other forces of nature. The theories proposed are called "Supergrand Unified Theories". They include quantum gravity, or "supergravity", and "supersymmetry", i.e. a symmetry between particles with integer spin (bosons), like the photon, the graviton and the particles W^+, W^-, Z^0 of the electroweak force, and particles of half-integer spin (fermions), like the quarks, the electron and the neutrino. The main prediction of the super-symmetric theories is the existence of a new set of "supersymmetric particles", like the photino, which corresponds to the photon, the gravitino, which corresponds to the graviton, the squark, which corresponds to the quark, and so on. If these particles are discovered, supersymmetry will be established. Up to now, however, these theories are still rather speculative.

8.8 The Inflationary Universe

In recent years a new theory for the Universe has been put forward in order to explain some of the basic problems of cosmology. This is the theory of the "inflationary" Universe, suggested in 1980 by *Kazanas* and, in a more complete form, by *Guth* (1981) and, independently, by *Sato* (1981). The theory explains: (1) why the Universe is so homogeneous and isotropic and (2) why the whole Universe is so close to the flat model (whether it is open or closed). Indeed, the Universe could have been very open, if its density were much less than the critical density ϱ_c (Sects. 6.5, 6.6), or very closed if its density were much higher than the critical density ϱ_c. As we saw in Chap. 6 the critical density distinguishes the open models, according to which the Universe is infinite, from the closed models, in which it is finite. The observed density of the Universe is estimated to be at least 1/10 of the critical one, but it could reach the critical value if the neutrinos (or other particles predicted by the modern unified theories) have masses significantly above zero.

If the expansion of the Universe indeed follows one of the simple expansion models (Fig. 6.1), then the density in the very early Universe was very close to the critical density. It is estimated that the divergence from the formula (6.37), during the early stages of the expansion, could not be more than $1:10^{49}$, i.e. extremely small. The question arises, therefore, as to why the density of the Universe is so close to the critical density, and not a million times larger, or smaller.

The inflationary theory answers this point by suggesting that the Universe passed through a phase of enormous expansion (inflation), when its age was between 10^{-35} sec and 10^{-32} sec. During this stage the radius of the Universe increased by a factor of the order of 10^{50}. Thus, a region smaller than a proton became much larger than the whole visible Universe today. In this way, any initial deviation from the critical density became very small, and the current density of the Universe must be very close to the critical one. That is, whether the Universe is infinite or finite, it must still be very close to flat.

This case is similar to that of an expanding balloon. Whatever the initial curvature of the balloon, after a long period of expansion it becomes increasingly flat, i.e. any small region approaches, asymptotically, a plane.

The observed homogeneity and isotropy of the Universe is due to this enormous inflationary expansion. Indeed, it is estimated that when the age of the Universe was $t = 10^{-35}$ sec, the Universe was made up of many independent regions, each with radius 10^{-25} cm. That is, the horizon of each region ($r = ct$ where $t = 10^{-35}$ sec and $c = 3 \times 10^{10}$ cm/sec, the speed of light) was a trillion times smaller than the radius of a proton. After the period of inflation, however, its radius increased until it was larger than the radius of the visible Universe. With such a dilation, any initial irregularities and inhomogeneities must have been smoothed out.

During the inflationary stage, the Universe expanded exponentially, as in the de Sitter model (Sect. 6.1).

The enormous energy required for the inflationary expansion is attributed to a transformation of the "vacuum". The idea that the vacuum is not simply an empty space, but has certain interesting properties, comes from Heisenberg's uncertainty principle. According to this principle, the energy E has fluctuations ΔE which can be arbitrarily large over a correspondingly small time interval $\Delta t (\Delta E \cdot \Delta t \simeq \hbar)$. If $\Delta E = 2mc^2$, where m is the mass of the electron, the energy ΔE may create an electron-positron pair, which may live for a time $\Delta t = 10^{-21}$ sec. Our current understanding of the vacuum is that it is full of such particle-antiparticle pairs, which are continuously being created and annihilated. These particles could be electrons and positrons, protons and antiprotons, or similar pairs, which are called "virtual", in contrast to ordinary particles which live for much longer intervals of time. The vacuum, therefore, is not really a vacuum, but is a sea of virtual particles.

The modern theories, however, accept that there is more than one kind of vacuum. Besides the ordinary, or "real", vacuum, there is also the "false vacuum". Such a false vacuum exists, for example, inside a proton. The quarks move freely inside it, but they are not allowed to go out and move in the real vacuum, due to the enormously strong forces (the strong nuclear forces) which hold them inside the proton. The false vacuum has a higher energy than the real vacuum, but it only exists inside the hadrons.

The current grand unified theories (GUT) accept that in the very early Universe, the strong nuclear force was as strong as the weak and the electromagnetic forces. At this stage, all the matter in the Universe was inside a false vacuum. This happened when the age of the Universe was less than 10^{-35} sec, and its temperature was more than 10^{27} K. This time is called the "grand unification time" (GUT time). After this time, the temperature dropped and the false vacuum became a real vacuum. The symmetry between the strong and the electroweak forces was broken. At the same time, an enormous amount of energy was released, the excess energy which the false vacuum held over the real vacuum, into which it was converted.

Guth calls this phenomenon a "phase transition", just like the transition from a liquid to a solid state. In such a transition, the structure of the molecules changes, and a significant amount of energy is released. In the case of the GUT theories, what changes structure is the vacuum, and the energy released causes the inflationary expansion of the Universe.

The energy density of the "false vacuum" can be considered as due to a large cosmological constant which acts to produce the exponential expansion. In fact, Einstein's field equations (6.3) can be written in the form

$$G_{\mu\nu} = -\lambda g_{\mu\nu} - \kappa T_{\mu\nu} \tag{8.26}$$

therefore, the energy density tensor $T_{\mu\nu}$ is supplemented by the term containing λ, which is the energy due to the (false) vacuum. This is most evident in the case of the de Sitter Universe (Sect. 6.1), which is empty, i.e. $T_{\mu\nu} = 0$, but has λ different from zero. This modern interpretation of Einstein's field equations allows the new ideas about the structure of the vacuum to be incorporated in the usual equations of general relativity.

The inflationary phase of the Universe lasts between 10^{-35} sec and 10^{-32} sec. During that period, the Universe passes through a phase similar to that generated when a liquid is "supercooled", during which time its energy density is constant. After 10^{-32} sec, the expansion of the Universe follows the law appropriate for the radiation era (8.4). This is an adiabatic expansion, without any external contribution of energy, which continues up to the matter era (Fig. 8.1), and after a slight modification, up to the present time.

The tremendous inflation of the Universe seems to solve one more cosmological problem associated with the grand unified theories. These theories predict the production of large numbers of magnetic monopoles. The theoretical density of monopoles is 10^{12} times larger than the observational limits found up till now. But if we accept inflation, we can explain the observed scarcity of monopoles, because inflation decreased their density by a much larger factor.

All these properties of the inflationary scenario make this theory very attractive. The original theory, however, had a fatal flaw that made it unacceptable. Namely, at the end of the inflationary period the state of the vacuum was assumed to be normal, and the value of the cosmological constant to become zero or very close to zero. This transition, however, introduced a great number of large inhomogeneities (bubbles), which destroyed the nice homogeneity and isotropy that had been achieved during the inflation period.

In order to remedy this defect, a "new inflationary theory" was proposed in 1982 by *Albrecht* and *Steinhardt* and independently by *Linde*. The improvement comes by making a smooth transition from the false vacuum to the real vacuum. In other words, the change of the cosmological constant λ from its high value during the inflationary period, to zero (or very small) afterwards, is not abrupt, as in the original theory, but gradual. In this way, the fluctuations introduced are not as large as in the original case. One may say that, while inflation eliminates any original fluctuations, the period immediately after inflation introduces new fluctuations. These fluctuations are supposed to act as seeds for the generation of structures in the Universe at later stages (galaxies, clusters and superclusters, Sect. 8.3).

But even the new inflationary theory has two important difficulties. The first difficulty is that its perturbations are at least 10^5 times larger than the observational limits set by the primordial microwave radiation. The second difficulty is that this theory does not lead, after the GUT time, to the expected separation of the strong force from the electroweak force (the electric and weak forces are still united up to 10^{-12} sec), but to an unphysical situation.

These difficulties have made necessary a new revision of the theory. The most modern version of the inflationary scenario has been proposed by *Nanopoulos* and his collaborators at CERN (1983). This theory is called "primordial inflation", because it assumes that inflation started well before the grand unification time, near the Planck time itself. In this way, the large fluctuations of the old theories are avoided. This "primordial inflation" comes to an end after the GUT time, as in the previous theories. Another improved version of the inflationary theory was proposed by *Linde* (1983). This is called "chaotic inflation" because it assumes that inflation started in a chaotic state near the Planck time.

However, if inflation started well before the GUT time, what caused it? The "natural" explanation based upon the transition of the "false vacuum" to the "real vacuum" is no longer available. Also, the inclusion of gravity is based upon specific theories of quantum gravity. No such theory, however, can be considered established. On the other hand there are specialists, like *Penrose,* that doubt all inflationary scenarios. Thus the problem of inflation in the Universe is still open.

9. The Evolution of the Universe

9.1 The Arrow of Time

It is obvious that time has a certain direction; it flows from the past to the future. The past is essentially different from the future. The past is already a fact, something which cannot be altered, while the future is something unertain and to a certain degree it depends on us. All events of everyday life have a certain direction, the direction of increasing time. This is particularly obvious if one projects a film backwards. The inverse sense of events is so unnatural, that it ends up being ridiculous. You can see the athletes running and jumping backwards, lifting themselves away from the ground to grasp the pole which is moving in their direction, or waiting until the javelin comes from the ground to place itself in their hands. If the time flows in the right direction, such strange events do not happen. What is most important is that the natural direction of time ensures that the cause precedes the effect. For example, the athlete throwing the javelin is the cause which comes before the effect which is the javelin striking the ground.

Although this seems so obvious, we notice that almost all the laws of nature are symmetric with respect to the direction of time. Let us consider a planet moving from a position A at time $t = 0$, to the positions B and C at times $t = 1$ and $t = 2$ respectively. If, at time $t = 2$, the velocity of the planet is exactly reversed, its orbit follows exactly the same steps C, B, A. This, however, is nothing but the original movement described backwards in time, from position C to B to A at times $t = 2$, $t = 1$ and $t = 0$ respectively. We also note that the physical laws do not change if the Earth or the Moon start spinning backwards. This backward spin is simply a forward spin with time reversed.

The same is true for the other forces of nature, the electromagnetic and the nuclear forces (apart from one exception we shall mention later on). Therefore, if we know the state of a system at a certain time, that is if we know the exact positions and velocities of all the particles in the system, and the forces which act upon them, we can predict its future and, equally accurately, calculate its past. For example, we can equally well compute lunar eclipses for the year 4500 A.D. as for the year 500 B.C.

Thus it seems that the physical laws, in general, do not recognise a direction of time. This is a problem. Why doesn't the basic concept of a direction of time appear in an obvious way into the physical laws? And what defines the sense of direction of time? This question is particularly relevant to cosmology which examines the evolution of the whole Universe. Is there a sense of time direction in the Universe? If yes, then it is meaningful to seek the cause of the Universe. If no, then instead of having a cause, we have a succession of various states of the Universe, each one of which can be considered as the cause of either the previous or the subsequent state.

Let us examine the subject more carefully, and let us see where the sense of the direction of time comes from. There are four types of phenomena where the direction of time is important:

a) The increase of entropy.
b) Psychological time.
c) The expansion of the Universe.
d) The time asymmetry in the decay of K^0 particles.

9.1.1 The Increase of Entropy

The increase of the entropy of a closed system is the most fundamental principle of thermodynamics, the famous "second law of thermodynamics". The entropy S is essentially the logarithm of the probability P for the system to be found in a given state:

$$S = k \log P, \tag{9.1}$$

where k is Boltzmann's constant. The second law of thermodynamics tells us that the probability and the entropy of a closed system increase continuously. If two bodies of different temperatures come into contact, it is immensely more probable that heat will pass from the hotter body to the cooler than the other way round. Also, if we put a sugar cube in our tea, the cube will soon dissolve and the tea will be uniformly sweet. The probability of finding the sugar molecules uniformly distributed in the tea is much higher than the probability of having them concentrated in a cube. That is why the dissolution of the cube is characterised by an increase in probability and entropy of the system.

The reverse sequence of events is extremely improbable. We cannot rule out that the randomly distributed molecules of sugar in the tea may gather together to form a cube, which, due to random pushings from the tea molecules, may be ejected from the teacup. This, however, is so highly unlikely that even if we had billions of cups of tea with sugar in front of us, and we waited for the whole age of the Universe, we would not see even one cube of sugar ejected from a cup.

We notice, therefore, that the normal direction of time is escorted by an increase in probability and entropy. If the direction of time were changed in some cases, the microscopic expression of the laws of nature would not change, but some macroscopic expression of these laws would, i.e. we would see sugar cubes coming out of teacups, athletes jumping backwards and so on.

As we saw, the direction of the normal flow of time is identical to the direction from cause to effect. That is, the cause always precedes the effect. In the reverse direction of time, we would still like to say that the cause precedes the effect. That is why we mentioned the motions of the tea molecules as the cause for the gathering of the sugar molecules into a cube and its ejection from the cup. In this way, however, we *do not explain* how the movements of the tea molecules were so well organised to conspire to produce such unlikely events. The explanation is very simple in the normal flow of time. The sugar did not enter the tea by itself, we put it in. The cause is outside the system of cup, tea and sugar. In the opposite flow of time, we do not have an external factor (the man) and that is why we find such strange effects.

If we ignore the direction of time and keep only the continuity of time, we may say that a certain state of a system is the cause for its next state, as well as the cause for its previous state. Then the usual concept of causality is lost. The shot from the murderer is the cause of the victim's death, but if the direction of time does not matter then the death of the victim is the cause of the murderer's shot too. The last statement, however, is so contrary to our experience that it cannot be accepted. We conclude, therefore, that the concept of increase in probability and entropy is directly related to the concept of causality. This, in turn, is due to the fact that the result is a situation more probable than the situation which is characterised by the cause.

A particular case where the increase in entropy has a special application in nature, is the emission of radiation. The radiation propagates away from the source towards other bodies. The light of the Sun, for example, illuminates the Earth or is dispersed into space. The reverse is extremely improbable. That is, when the Sun is extinguished, the light cannot be gathered again from the depths of space to light the Sun again. In the language of science, we say that the electromagnetic radiation is due to a "retarded" potential. This means that the radiation gives us information about a change in the source, after some delay. The reverse, an "advanced" potential, is not excluded by the equations of electromagnetism, but is excluded by the second law of thermodynamics.

9.1.2 Psychological Time

Clearly, the impression humans have that time flows continuously is due, at least partly, to the biological functions of their bodies. Most biological functions increase the entropy of the organism and cause its evolution, and eventually its death. There are, however, functions of "growth", particularly during childhood, which cause an increase in the organisation of the organism. (Even so, the total entropy of the system, which consists of the person and his environment, always increases.)

It is also interesting to examine the relation between the mental functions of a person and entropy. The gathering and organisation of information by the human brain is something which at first glance contradicts the notion of the increase in entropy. In this case again, however, the total entropy of the human brain and its environment must increase. There is a branch of mathematics, information theory, which studies the gathering, correlation and organisation of information. This theory uses the concept of "negative entropy" to characterise this organisation.

In psychological time, there is also another element which distinguishes the past from the future, and that is memory. We have lived through the past, its traces are left inside us. Therefore we cannot change it, while we always have the impression that the future has some element of uncertainty, that we can affect it at least partly. This impression is related to the freedom of will in man. If a person does everything in a fully predetermined way, then his future is absolutely certain and depends only on the past. But then, if the future is fully determined, there is nothing to distinguish a true syllogism from a false one, since anything a man may think is necessarily as it is and not otherwise. Therefore, neither a man's logic nor his science have any value, because they cannot be different from what they are.

The problem of the relationship between consciousness and freedom of will on the one hand, and the physical environment extending up to the whole Universe is particularly complex and we are not going to enter into it here. However, in Sect. 11.5 we will consider the possibility of a deeper relation between man and the Universe. Namely, it is possible that the existence of man is related to the structure of the whole Universe.

9.1.3 The Expansion of the Universe

We observe that the Universe expands in the normal direction of time. Thus, we could *define* as the positive direction of time, the direction in which the Universe expands. What, however, is the relationship between the expansion of the Universe and entropy? If the Universe starts contracting one day, will its entropy start decreasing? This seems highly unlikely. Indeed, the course of events on the Earth does not seem to have any relevance to whether the Universe expands or not. If the expansion stops at some stage and the Universe starts to contract, it seems inconceivable that the flow of time will change direction. If this could happen, the light would start returning to the stars instead of going away from them and so on, something which is entirely unlikely. It is much more likely that the direction of time in a contracting Universe is the same as in an expanding one.

The approach to the "final singularity" is particularly interesting, since the whole Universe will have collapsed then. This process has been examined by *Penrose* in an interesting paper (1979). If the direction of time were to change during the collapse of the Universe, then the final collapse would be nothing but a reflection, as if in a mirror, of the Big Bang. But if we accept, as is more probable, that the contraction of the Universe does not change the direction of the increase in entropy, then the contraction of the Universe will result in a significant increase in the entropy and the state of the final collapse will be significantly different from the initial state of the Big Bang.

The increase of the entropy is particularly relevant to the collapse of a black hole. As *Bekenstein* and *Hawking* calculated, a black hole has an entropy which is proportional to its area (by area we mean the area of the surface of the horizon, i.e. a sphere of radius equal to the Schwarzschild radius of the black hole). The fall of matter inside the black hole increases its mass, therefore its horizon radius and thus its entropy. Therefore, the collapse into a black hole increases the entropy instead of decreasing it. *Penrose* attributes this to the peculiar nature of gravity.

Entropy of Gravitating Systems. Gravity is distinguished from all the other forces of nature because it is always attractive, and when two bodies approach, the gravitational forces increase indefinitely. The result is that gravitating systems become clumpy while other types of forces (like electromagnetic forces, which are partly repulsive) tend to diffuse matter. This peculiar behaviour of gravity becomes clear if we consider a satellite orbiting the Earth. If we decrease the satellite's velocity, it will come closer to the Earth, and because the Earth's attraction will increase, the satellite will move *faster*. Also, if we subtract heat from a stellar system, that is, if we reduce the random motions of the stars, the whole system will contract with the result that the random motions of the stars will increase and thus the "temperature" of the system will increase as well. In general, the increase in the entropy of a system means that the system becomes denser.

This behaviour of gravity is different from the dissolution of our sugar cube in the tea, which is due to electromagnetic forces. In the case of the tea, the sugar tends to dissolve as uniformly as possible. In the case of a gravitating system, the tendency is exactly the reverse: condensations tend to form. Thus the segregation of matter into galaxies and stars in the expanding Universe, is accompanied by an increase in entropy of the Universe. Also the creation of black holes implies an increase in entropy. In this way, the entropy of the final contraction of the Universe will be very high, while the initial entropy was relatively low.

Penrose discusses these arguments in conjunction with suggestions by other researchers who remark that during the first stages of the expansion of the Universe the entropy was "high". A measure of the entropy during the initial stages of the expansion was the number of photons per baryon, which was 10^9, i.e. quite high. Nevertheless, according to *Penrose*, the entropy of the enormous black hole resulting from the collapse of the Universe will be of the order of 10^{40}, that is incomparably higher than the entropy during the first stages of the Big Bang.

A consequence of *Penrose*'s ideas is that there are no "white holes" in the Universe (unless we call the Big Bang itself a white hole). Indeed, if white holes existed, their behaviour would be different from the behaviour of black holes, i.e. their entropy would tend to decrease. Further, the appearance of a white hole in the Universe is without a cause, i.e. it cannot be predicted, unlike the appearance of a black hole. Therefore, the asymmetry in the behaviour of time is reflected in the asymmetry of the singularities in spacetime. The Universe creates black holes but no white holes.

9.1.4 The Decay of K^0 Particles [1]

Until a few years ago it was thought that all laws of microphysics were symmetric with respect to time. This was known as *T* symmetry. An inversion of the direction of time was not thought to lead to a different result. At the same time, a symmetry (C) with respect to charge was observed (e.g. symmetry between the production of protons and antiprotons or electrons and positrons during nuclear reactions). There was also symmetry between left-handed and right-handed particles, called P symmetry, or parity. However, *Yang* and *Lee* (Nobel prize 1957) showed that P symmetry is not always preserved. They showed that during certain reactions more right-handed electrons are produced than left-handed ones. Similarly, it was found that charge symmetry C does not always hold exactly. Thus the two symmetries were replaced by one, the CP symmetry. That is, it was accepted that the production of a certain number of right-handed particles was accompanied by the production of an equal number of left-handed antipraticles. However, *Cronin* and *Fitch* (Nobel prize 1980) showed, in 1964, that even this symmetry did not always exist. They came to this conclusion after observing the decay of a K^0 particle to pions. If CP symmetry existed, this decay would not be possible. *Cronin* and *Fitch* observed several forbidden decays, so that the lack of CP symmetry is now well established.

If CP symmetry does not always exist, then T symmetry cannot exist either. Indeed, there are several reasons, both experimental and theoretical, which confirm that CPT symmetry exists (at least it exists to a very high accuracy). That is, if in the normal direction of time more right-handed particles are produced than left-handed antiparticles, the opposite

---

[1] The K^0 particles are called kaons; they are mesons with a mass approximately equal to half the proton mass. They are particularly short lived.

is true if time is reversed. Therefore, the experiments of *Cronin* and *Fitch* showed that there is a basic time asymmetry in the fundamental laws of microphysics.

At first sight this asymmetry seems to be small (1 in 500) and does not seem to have any relation to the rest of the phenomena where the asymmetry with respect to time is observed, i.e. the increase of entropy or the expansion of the Universe. During the first stages of the expansion of the Universe, however, this basic asymmetry seems to have played a very important role. As *A. Sakharov* has shown, the creation of more matter than antimatter in the Universe is characterised by a certain "direction of time", which is related to the CP asymmetry. We again see, therefore, that the "direction of time" is one of the fundamental characteristics of the Universe.

9.2 The Future of the Universe

We have, up to now, studied the current state of the Universe and its history right from the beginning of the expansion. The problem of the future evolution of the Universe has been studied to a much lesser extent by scientists. This is mainly due to the fact that one cannot verify any predictions for the end of the Universe. Besides, if we make assumptions about the early Universe, we always expect to see some consequences of these assumptions today. This, however, is impossible for the future.

Even so, it is possible to calculate the future of the Universe if we make the basic, but plausible, assumption that the laws of nature are known. The general theory of relativity is particularly important in this sort of calculation.

There are two basic scenarios which describe the future evolution of the Universe. The first refers to a closed, pulsating Universe and the second to an open, continuously expanding Universe.

As we saw in Sect. 6.6 both scenarios are possible in the actual Universe, so we discuss them in turn.

9.2.1 Pulsating Universe

In this case the expansion of the Universe will eventually stop and contraction will commence. The galaxies will start to approach each other, while the temperature of the background radiation will continuously increase. During the contraction, galaxies will collide and a population of high velocity stars will form. Subsequently, the stars will be destroyed either by collisions between themselves, or (mainly) because of the effect of the background radiation which will then be very intense. The evaporation of stars will create a diffuse gas at very high temperature. The temperature will become so high that the atoms and nuclei will dissociate into elementary particles, which in turn will decay into photons and a quark-gluon plasma, so that a new radiation dominated era will start, just as in the first stages of the cosmic expansion. The contraction proceeds until the Universe collapses to a singularity in spacetime, as at the beginning. This is called the Big Crunch.

Some people assume that after the condensation of the Universe into a "final" singularity, a new phase of expansion will start which will be followed by a new contraction and so on. So they speak about successive pulsations of the Universe. However this theory suffers from two basic difficulties. The first difficulty is that we cannot consider the new Universe as causally related to the old one. Causality cannot be continued through a singularity of infinite density. We have already encountered this difficulty on a smaller scale when we considered the possibility of transforming a black hole into a white hole. This difficulty is much stronger here because the Big Crunch absorbs the whole Universe into a singularity. Therefore, we cannot bypass the singularity and reach a new state which is causally related to the old Universe. Everything has to pass through the melting pot of the Big Crunch.

The second difficulty of a recycling Universe is related to the increase of entropy. As we saw in Sect. 9.1, the entropy of the Universe increases continuously, and for this reason its "final" form will be essentially different from its initial form. The "final" entropy of the Univese will be enormous, while its "initial" entropy was relatively low. The distinction between the beginning and the end of the Universe has been particularly stressed by *Penrose*, who has calculated quantitatively the increase in entropy of the Universe from its first stages, when its entropy was less than 10^9 (and possibly zero at $t = 0$), to the final value of 10^{40} in the final collapse of the Universe. *Penrose's* ideas have become increasingly accepted today. The conclusion is that the final singularity of the Universe will be entirely different from the initial singularity.

Therefore, the contraction of the Universe is not symmetric to its expansion and thus we cannot say that the "final" singularity can be the beginning of a new Universe which will have roughly the same history as our own.

One consequence of such a collapse will be the disappearance of any trace of life in the Universe. That is, a closed Universe leads inexorably to the end of life, and in particular the end of intelligent life.

A way to avoid this prospect is by allowing only one part of the Universe to collapse, by forming lots of black holes, while the rest would start expanding again. It seems, however, that the theorem by *Hawking* and *Penrose* does not allow any part of the Universe to avoid collapse. In any case, even if such a phenomenon could happen, it would probably come too late to save life. This is because the enormous temperature of the gas and radiation during the final stages of the collapse would destroy any form of life and any organic molecules.

9.2.2 Continuously Expanding Universe

Let us consider now the second scenario which describes an infinite and continuously expanding Universe. In this case we have a continuous expansion and disintegration of the Universe. The stars will exhaust all their energy reserves and will either end up as white dwarfs (which will eventually become black dwarfs and will be extinguished), neutron stars or black holes. The time required for all stars to be put out is of the order of 10^{14} years (a hundred trillion years). It is estimated that the planetary systems will take about the same time to be disrupted by the tidal effects of neighbouring stars. The galaxies are expected to disintegrate in about 10^{19} years. At the end, they will comprise only a binary star, or a stable multiple system of stars, formed in their centres, while most single stars will have escaped. In any case, the stars, either single or members of multiple systems, will be dead bodies emitting no radiation. Later on, even black holes will radiate away their contents (by the *Hawking* effect) and disappear.

There will still be solid bodies in the Universe, like the interstellar dust and the planets, as well as diffuse gas, which, because of the expansion of the Universe, will become increasingly dispersed. Finally, the radiation from the stars will spread out uniformly through space and, together with the microwave background radiation, will become more and more diffuse, with a temperature approaching absolute zero.

This future appears just as disappointing for life and man as the prospects in the case of a pulsating Universe. Life does not have much choice between being baked at infinite temperature or frozen at absolute zero. In the second case, however, the end comes so slowly that various possibilities have been discussed so as to avoid this gloomy prospect.

Dyson made some interesting speculations concerning the prospects for life in an expanding Universe. His basic assumption is that intelligent life is not characterised by the type of matter from which it is made, but by its structure. He notices that if we assume that organic substances, with their known properties, are necessary for the existence of life, then life will finish when all the sources of energy which can maintain a temperature high enough to sustain metabolic rates will be exhausted. If, however, we consider that the future evolution of life can take other forms, entirely different from the present ones (e.g. electronic), it is not out of the question that life may be preserved forever. At this point we are bordering upon science fiction where imagination prevails. However, imagination is a plus in such daring extrapolations of logic. That is why *Dyson* refers to novels such as "The Black Cloud" by *Hoyle*. The "black cloud" is an interstellar cloud with life and intelligence consisting of dispersed molecules, which interact with each other via electromagnetic waves. If such possibilities exist, it is not out of the question for life to be preserved close to absolute zero in billions of years from now. Indeed, the temperatures of interstellar clouds are barely a few dozen degrees above absolute zero. According to *Dyson*, life in such forms can be preserved indefinitely.

Dyson talks about time spans of 10^{76} years! Over such periods of time many things can happen. There is, however, an insuperable obstacle for the realisation of *Dyson*'s prospects. It comes from the current grand unified theories of physics. If these theories are correct, the proton is unstable with a lifetime of the order of 10^{31} years. Therefore, the matter in the dead stars, as well as the interstellar and intergalactic matter, will decay away over periods longer than this. No elementary particles like protons and neutrons, of which matter is largely made, will remain. Therefore, neither atoms or molecules, nor solids or liquids out of which life may be formed, will exist. This is because it seems impossible to construct composite bodies, and finally life, out of only electrons, photons and neutrinos.

Not even electrons can be preserved forever. They will annihilate with positrons (8.17) to create photons. Indeed, electrons and positrons do form pairs which resemble hydrogen atoms. Such a pair is called "positronium", but it is unstable[2]. That is, the two particles eventually approach each other and annihilate. This phenomenon can be observed in our laboratories. A recent calculation, however, (1982) has shown that after the protons and neutrons have decayed, the positronia will have enormous dimensions so that their annihilation will take a very long time. The formation of these pairs will take place when the age of the Universe will be 10^{71} years (because before this time the random motions of the electrons and positrons will not allow them to combine into pairs). The pairs which will form then will have dimensions as large as the present separation of the Sun from the Earth. In this way, the gradual approach of the two particles, leading to annihilation, may last for 10^{116} years. After this time neither hadrons nor electrons will exist. The Universe will be a cold gas of photons and neutrinos (with a temperature of 10^{-70} K or less) which will tend towards complete disintegration.

There does not, therefore, seem to be any way to avoid the conclusion that life is doomed to extinction in the far future, in a continuously expanding Universe. It is very characteristic of this exercise in extrapolation that the suggestions of *Dyson* (1979) were so quickly overturned by the theory of the proton decay which seems so plausible today.

Of course, we cannot exclude the possibility that our theories may change in the future. For example, it is not impossible that the cosmological constant is different from zero, in which case other cosmological models may describe the Universe. It seems, however, that the two basic scenarios described above are unlikely to change. That is, the Universe will either suffer a final collapse or it will expand forever. In both cases, its entropy will increase continuously, and any form of organisation, such is life, will be destroyed.

---

[2] There are two types of positronium, with lifetimes of 10^{-7} and 10^{-10} sec respectively.

Part III

Fundamental Problems

10. The Physics of the Universe

10.1 The Unity of Nature

Cosmology is unique among sciences because it deals with the whole physical Universe, while other sciences deal with only part of it, or with certain aspects of physical reality.

This uniqueness of cosmology makes it particularly interesting and also particularly difficult. All fundamental and unsolved problems of the other sciences usually lead us to cosmological problems. For example, chemistry studies the properties of the chemical elements; but why do these chemical elements exist on the Earth? The problem of the formation of the elements leads us to the initial creation of the Universe or to the formation and evolution of stars, that is, directly to cosmology or to branches of astrophysics intimately related to cosmology.

Similarly, the existence of a solid crust on the Earth, the existence of oceans and even more, the existence of life on the Earth are directly related to the conditions under which the Earth and the whole solar system were formed. As we shall see later, small differences in the initial conditions and properties of the Universe could lead to a completely different Universe, without stars, without planets and without life.

Another fundamental question also closely relates cosmology to physics and the other sciences. This refers to the unification of the physical laws, and their reduction to other more fundamental laws. The unification of the forces of nature, about which we talked in Sect. 8.5, is directly related to cosmology.

This unification is present in various branches of science. For example, the complex movements of the celestial bodies led *Newton* to the universal law of gravity which can explain all relevant phenomena, apart from some corrections brought about by the general theory of relativity. *Einstein*'s theory achieved an even deeper unification by treating gravity as a geometrical property of spacetime. A similar revolution came about in chemistry and biochemistry, when all forces relevant to the chemical phenomena were shown to be of electromagnetic nature. Today, physicists go even further in the same direction, in trying to unify all the forces of nature, from the nuclear to the electromagnetic and gravitational forces.

At this point, however, a basic and fundamental question arises. Is it possible to show why the laws of nature are what we observe and not different? For example, can we deduce that the law of gravity must be the way it is (namely an attractive force between two bodies proportional to their masses and inversely proportional to the square of their separation)? At this point people disagree. Some scientists, like *Einstein* and *Eddington* think that the basic laws of physics can be deduced by syllogisms a priori. Most scientists, however, accept experience as the only starting point for the expression of physical laws and their applications.

Of course, everybody accepts that observation and experiment provide the essential checks for the various theories. A *simple* theory, however, is much more attractive than any observation. That is why, when in 1919 the deflection of light due to the gravitational field of the Sun was measured and *Einstein* was asked what he would have said if the result were different from his prediction, he answered: "I would have said pity to God, because the theory *is* correct".

This incident shows *Einstein*'s confidence in his theory. It is true that *Einstein* was proved right not only on this occasion, but on many other occasions as well, when his theory was put to the test. However, very few people have the amazing intuition that *Einstein* had, and even *Einstein* himself failed when he tried to advance beyond general relativity by developing his Unified Theory. Therefore, not everything that looks simple is necessarily true. In some strange way, however, whatever is true, is simple, provided one looks at it from the right angle. For example the motions of the planets appear very complicated. *Ptolemy*'s attempts to explain them with his epicyclic theory failed. Even *Copernicus,* in spite of his basic simple idea that the Sun was at the centre of the solar system, needed even more epicycles than *Ptolemy* to explain the planetary motions. The problem was made simpler by the introduction of *Kepler*'s elliptical orbits, but its final solution was due to *Newton*. *Newton* found that the gravity law, expressed so simply, can account for all the complications in the motions of the planets.

Later on, when the general theory of relativity was introduced, it was thought that the simplicity of *Newton*'s laws was lost. This impression, however, lasted only as long as it took scientists to understand the new theory. The general theory of relativity not only explains all phenomena which *Newton*'s theory could explain and the deviations which *Newton*'s theory could not explain, but it also sheds light on the fundamental properties of gravity in a simple and elegant way. That is, *Einstein*'s theory *explains* why *Newton*'s laws hold (approximately). Thus, we find again the simplicity of the physical laws, at a level deeper than the level at which *Newton*'s laws apply.

The modern unified theories of the four basic forces of nature try to *explain* not only gravity, but also electromagnetism and the weak and strong nuclear forces. That is, they try to formulate some very general principles from which one can deduce all the complicated properties of matter in any form it may be, from the atomic nuclei to the galaxies and the Universe itself.

As a foundation for all these efforts, and a necessary condition for making deductions about the *simplicity* of the laws of nature, there is one preassumption which we often forget to mention. It is the *universality* of the physical laws. The same laws which hold on the Earth, hold on the Sun, in the Galaxy and in the most remote parts of the Universe as well. This idea which seems so obvious to us today, took centuries to become accepted. In the past centuries it was considered plausible that different physical laws might hold in different parts of the Earth, let alone in different stars of the sky.

It was the contradiction of this idea by many observations, which we shall discuss in the next section, that led to the gradual acceptance of the universality of the physical laws. Only when this universality became an established belief did it become possible for the big jump, to start talking about laws that govern the whole Universe. It became clear that the Universe has an amazing *unity*. This unity is a prerequisite for any other discussion of its *simplicity,* or for the *unification* of its laws.

The unity of the Universe is related to its common origin. Of course it is expected that the physical laws which hold in a galaxy will also hold in the stars which resulted from the fragmentation of this galaxy. In a similar way all the galaxies which started from the same initial explosion of the Universe, are governed by the same basic laws, the "laws of the Universe".

10.2 The Laws of the Universe

Our assumptions and calculations concerning the whole Universe, its past and its future, are based on two fundamental principles:

1. The physical laws are known. That is, we know which laws hold at all points in the Universe, and how they change with time, if they change at all.
2. The uncertainty in the physical laws does not affect our conclusions.

The first principle assumes that there are physical laws which hold at all points in the Universe and that our experience from the Earth can give us information about them. This assumption is based on the observation that the physical laws we find on the Earth apply up to the limits of the observable Universe.

The most important role in the establishment of this idea has been played by spectroscopy. Indeed, the properties of the atoms of matter as revealed by their spectra, seem to be the same whether we observe them in the laboratory or in distant galaxies. We find, for example, that the lines of the hydrogen atom are given by a very simple formula:

$$\frac{1}{\lambda} = R\left(\frac{1}{m^2} - \frac{1}{n^2}\right), \tag{10.1}$$

where λ is the wavelength of the line, m and n are integers and R is a universal constant. The same formula holds for the most distant objects in the Universe, the quasars which are billions of light years away from us. Spectroscopy proves also that the same chemical elements which we find on the Earth exist everywhere in the Universe and in similar proportions, and that the atoms and molecules have everywhere the same properties.

Further, the speed of light and the most important constants of physics, like Planck's constant, the electronic charge etc., are not only constant everywhere in the Universe, but they are also constant in time. Various observations of distant galaxies and quasars show that these constants change by less than $10^{-11} - 10^{-14}$ per year, and most likely they do not change at all.

The observations of the motions of distant binary stars show that Newton's gravity law is the same everywhere, while recent observations of distant "gravitational lenses" show that the deviations from Newton's laws, due to relativistic effects, are universal too. Similar observations hold for the magnetic phenomena in the Universe, as revealed by the Zeeman effect (the splitting of the spectral lines due to the presence of a magnetic field), and the synchrotron radiation (radiation from relativistic electrons moving inside magnetic fields). Therefore all electromagnetic phenomena follow the same laws at all points in the Universe. The same seems to hold for the nuclear reactions which produce the energy in the nearby and distant stars, and which are governed by the weak and strong nuclear forces. Generally, every observation concerning distant objects in the Universe reinforces our idea of the universality of the known physical laws. More evidence comes from our observations of the homogeneity and isotropy of the Universe and particularly our observations of the isotropy of the microwave background.

Our observations of distant galaxies, however, reveal how the Universe was in the distant past, when the light set off from those galaxies in order to reach the Earth, namely billions of years ago. Therefore, we have good indications that the laws of nature are not only universal, but that in general they do not change with time over a period of billions of years.

After this observation we do not hesitate to apply the known physical laws to the whole of the Universe, and what is more, to stretch them to the distant future and to the distant past as far as the beginning of time.

It is obvious that if this principle did not hold and the physical laws were different in every star or in every galaxy, it would not be possible to conclude anything about the Universe and its evolution. The whole of cosmology is based exactly on the principle that the laws of nature are known, or that it is possible for them to become known. This principle is particularly relevant to the current research concerning the properties of the Universe during the first fraction of the first second of its existence.

In certain cases the laws of nature show some *change* with time, but this change is, or is supposed to be, *known*. That is, the expression of the particular law is a function of the age of the Universe. For example, we may say that the density ϱ of matter in the Universe is everywhere the same, but continuously decreases as the Universe expands with a scale factor R which is a known function of time. Similarly, theories which assume a varying G give the law of this variation.

The fact that we have various theories concerning the Universe means that the physical laws are not entirely known. The expansion, however, of astronomical observations in space and the improvement of our experimental devices help us to reject one by one the inaccurate theories and to come up with more accurate ones which explain more and more phenomena in the Universe.

In spite of all this, there is no doubt that the extrapolation of our physical theories to conditions so different from those we can observe around us, like the conditions prevailing in the first moments after the creation of the Universe, is so vast that we cannot exclude errors. Therefore, the progress in cosmology is always made with restrained optimism for its future.

10.3 The Uncertainty of the Physical Laws

Let us examine now what uncertainties exist in the physical laws which govern the Universe. Such uncertainties obviously affect our possible predictions and conclusions. There are various sources for our uncertainty:

1. The uncertainty in our knowledge of the physical laws.
2. The various instabilities in the evolution of classical dynamical systems.
3. The finite velocity of light.
4. The uncertainty of quantum mechanics and possible human intervention.

Let us examine one by one the above causes:

10.3.1 The Uncertainty in the Knowledge of the Physical Laws

The first cause of uncertainty is obvious. For example, we know that Newton's gravity law describes the motions of the planets very accurately. If, however, we want to apply Newton's laws over a time span of billions of years in order to find out how the solar system was formed, we are met with insuperable difficulties. For a start, we know that we must take into account relativistic corrections, but by how much? The corrections we usually make are of the first order and can be easily applied. The higher order corrections, however, and particularly those which are due to gravitational radiation are not accurately known. Even if we omit these corrections as too small, we are still left with other effects which alter the orbits of the planets and which cannot be calculated easily. Such effects are mostly due to the interplanetary matter which exists everywhere in the solar system and acts as a brake on the motions of the planets. These effects can only be taken into account approximately but their results are essentially unpredictable over large timespans. It is particularly difficult to calculate the effect on the protoplanets (which were much larger than the planets are now) of the interplanetary matter which existed in the past and which has subsequently been dispersed into outer space, or accreted by the Sun.

Therefore, even if we use a very powerful computer and compute the orbits of the planets backwards in time by solving either Newton's or Einstein's equations, our results will not be particularly valuable if our calculations refer to many millions or billions of years.

10.3.2 Instabilities

If we knew all the laws of the Universe and the initial conditions, that is if we knew the forces which acted between the various particles in the Universe as well as their present positions and velocities, we would be able to calculate the evolution of the Universe both backwards and forwards in time, with as much accuracy as we require. This statement expresses the belief of complete determinism. *Laplace* (1814) expressed such a version of absolute determinism by saying: "An intelligence, which would know all the forces which act in nature and the relative positions of the beings which make it up at a certain time, would include in the same formula the motions of the largest bodies in the Universe and those of the lightest atoms, if it were large enough to be able to subject all these data to mathematical analysis. Nothing would be uncertain for it; the future and the past would lie in front of its eyes. The human mind with the perfection it achieved in astronomy gives only a weak picture of this intelligence".

Nowadays we have the possibility to somehow approximate the abilities of *Laplace*'s hypothetical intelligence, with the use of electronic computers. Indeed, we can achieve amazing accuracy with computers. Thus we can calculate astronomical events such as eclipses which happened thousands of years ago, or will happen in thousands of years time, with the accuracy of a second. We can even compute the motions in stellar systems which are made up of millions of stars.

However our predictions fail when our systems become very unstable. We refer here to instabilities of the solutions of various differential equations which determine the motions of the classical (not quantised) systems.

This uncertainty is well known to those who deal with numerical solutions of dynamical systems. For example, one may compute two stellar orbits which may be very close to each other to begin with, but which diverge from each other very rapidly. The distance between the two orbits increases exponentially with time, being equal to $\varepsilon\, 10^{at}$, where ε is the initial distance and a is some positive constant. This can be proved rigorously in many cases, e.g. the case of unstable periodic orbits. If $a = 1$, an initially insignificant separation will become 10^{10} times greater at time $t = 10$.

The quantity ε is essentially the "step" by which we compute the orbit. By increasing the accuracy of the computation we do not achieve much. If, for example, we make ε ten times smaller, $\varepsilon' = \varepsilon/10$, the distance between the two orbits will be equal to 10^{10} after $t = 11$ instead of $t = 10$, i.e. only a little later than previously.

If we wish to compute the orbit for much longer time intervals, say for $t = 10^9$, we would need a step $\varepsilon = 10^{-10^{10}}$, that is $10^{10\,000\,000\,000}$ times smaller. The computer we would need in that case would be bigger than the whole visible Universe, in order to compute the orbits with the required accuracy.

Therefore, to compute with sufficient accuracy the evolution of the visible Universe, we would need a computer bigger than the visible Universe. The implication is that no technical advance in the future will allow us to approach, as close as we would wish, to *Laplace's* hypothetical "intelligence". There is, therefore, an insuperable uncertainty which can be overcome only by a supernatural Being.

Let us now examine more carefully what is the practical significance of our inability to compute exactly the evolution of the Universe.

We know that there are statistical phenomena in nature which are described by statistical laws. Branches of physics, such as thermodynamics are based upon such statistical laws. In such cases we make statistical predictions which are based upon the concept of probability. It is particularly characteristic of statistical laws that while the probabilities may be exceedingly small for certain phenomena, they approach certainty for others. For example, consider a room containing about 10^{30} gas atoms and one atom setting off from the wall on the right. A few moments later it is equally likely for this atom to be in the right half or the left half of the room. We cannot, therefore, make a prediction as to where the atom will be at this time. On the other hand, we may say with certainty that at any moment almost half of the atoms will be in the right half of the room and half will be in the left[1].

---

[1] More accurately, the probability that there are 1% more atoms in the right half of the room, than in the left, is vanishingly small $1:10^{10^{26}}$, i.e. $1:10^{100\,000\,000\,000\,000\,000\,000\,000\,000\,000}$.

Therefore, although we cannot predict the behaviour of individual gas particles, we can predict certain basic properties of the gas as a whole. Such observations have been made since the time statistical physics was developed. The new element that has been added in recent years is that almost all deterministic systems include unstable orbits, and therefore the laws which govern these systems are, in fact, statistical.

This observation is particularly relevant to the classical problem of "N bodies" in mechanics and astronomy. For example, by studying the evolution of a stellar system we can say how many stars will escape over a long period of time (a few billion years), but it does not allow us to say which stars will escape (apart from certain special cases where the escape of a specific star can be checked easily).

10.3.3 The Finite Speed of Light

Up to now we have been discussing classical systems, without considering the effects due to the finite speed of light. We know, however, from relativity, that it is impossible for any interaction to propagate faster than the speed of light. It is therefore impossible to know the condition of the Universe at *this very moment,* in order to predict its future with certainty; we cannot even do it for our own future. Suppose, for example, that there are two stars at a distance of ten lightyears from us. One star is made of matter and the other of antimatter, and both are too faint to be visible to us. If today they should collide and annihilate in such a way that the resulting radiation is beamed towards the Earth, the Earth will be destroyed in ten years time; yet it is impossible for us to predict this event.

The possibility of predicting the future in general relativity is a more complicated problem because of the curvature of spacetime. The basic question is whether the knowledge of our past (including all of the information that has reached us since the beginning of the Universe, with velocity less than or equal to the speed of light) is sufficient to predict the future, our own future and that of the whole Universe. The answer is negative. This information is not, in general (i.e. in most cosmological models), sufficient to predict the future of the Universe, not even our own future.

A related problem in the theory of general relativity is the so-called "Cauchy problem". The problem is to predict the future of the whole Universe if we know the conditions along a space-like surface that extends throughout the Universe (finite or infinite). In many models of the Universe this problem has no solution. Therefore, even a knowledge of the whole Universe at a certain moment is not sufficient to predict its future. Our only escape is to try to avoid all such models of the Universe.

10.3.4 Heisenberg's Principle of Uncertainty and the Uncertainty of Human Actions

On top of the various uncertainties we have already mentioned, we must mention two more which prevent us from fully predicting the future. They are *Heisenberg*'s uncertainty principle of quantum mechanics and the uncertainty involved in human actions. Both of them may be very important in certain cases. Let us consider the following hypothetical situation: A source emits electrons which, because of *Heisenberg*'s uncertainty principle have equal chances to hit or miss a target. If the target is connected to an atom bomb then the chances are half and half that the bomb will explode when one electron is emitted.

Equally uncertain is the future of the Earth now that people have sufficient nuclear weapons to destroy it entirely. If an astronaut travels for 1000 Earth years at a speed approaching that of light he will still be young when he returns to Earth 1000 years later. The question is, however, whether the Earth will still be here then or whether it will have been destroyed. Of course, the uncertainty is much greater when we consider longer periods of time.

The two classes of phenomena (quantum indeterminacy and indeterminacy of human actions) may be closely related. In fact human free will is considered by many people as based on the indeterminacy of quantum mechanics. This does not mean that human actions are arbitrary, but only that they are not completely determined causally as the usual grand scale phenomena of physics are.[2]

The indeterminacy of quantum mechanics has been considered by many physicists as a stumbling block that will eventually be eliminated from physics. The fact that we cannot simultaneously measure the position and velocity of an electron is supposed to be due to the imperfection of our measuring systems or methods, and it would be difficult to accept that an electron does not possess both a position and velocity at any given time. According to this point of view the indeterminacy of quantum mechanics is not essentially different from the uncertainties of classical mechanics. Quantum mechanics, in the same way as statistical mechanics, is a theory which bypasses our lack of exact knowledge by considering large numbers of particles, or waves. If we knew all the "hidden parameters" of the elementary particles we would have an equally deterministic system as in the case of classical mechanics, despite its practical difficulties.

---

[2] Although we cannot discuss here the problem of free will, we only point out that if human actions were completely determined, there would be no way to distinguish between truth and falsehood, between science and nonsense. Because everything we say, or think, would then be absolutely determined. Eddington gives the following nice example. Suppose that our brain produces sugar if we say that 7 times 8 equals 56, while it produces chalk if we say, wrongly, that 7 times 8 equals 65. We cannot say, however, that sugar is "better" or "more correct" than chalk.

This point of view, however, tends to be rejected by recent experiments. In fact it has been shown by *J.S. Bell* that if quantum phenomena were determined by some "hidden parameters", this would lead to certain experimental consequences that would be different from those of quantum mechanics. Certain observable quantities would be smaller in the classical case of hidden variables than in the corresponding quantum case. These are the famous "Bell inequalities" that have been tested by many experimental physicists in recent years. All the experiments show unambiguously that the quantum mechanical picture is correct, and thus the theory of hidden variables tends to be discarded. It seems that quantum indeterminacy is an objective phenomenon and not merely a lack of knowledge. After all, no one knows the exact nature of an electron, or of any other elementary particle, in order to ascertain that they should behave in the same way as the macroscopic bodies of our everyday experience.

Besides the quantum indeterminacy, it was pointed out by *Hawking* that the radiation of black holes has one further degree of uncertainty. In the usual quantum phenomena we can accurately determine one of two conjugate quantities (position or momentum) while the other is indeterminate, but in the case of particles emitted from a black hole both quantities are unknown. Thus a black hole introduces a larger indeterminacy than quantum mechanics. These recent developments have a great theoretical interest.

However, the quantum uncertainty in subatomic phenomena is negligible on the cosmic scale and human behaviour, no matter how irrational it becomes, is unlikely to change the course of the whole Universe, even millions of years hence, when man will have colonised the whole Galaxy.

The basic conclusion from our examination of the uncertainty in the physical laws is that certain phenomena cannot be calculated accurately. For this reason, in cosmology as in physics, we use two kinds of laws; the "exact" ones which describe the large scale phenomena, such as the expansion of the Universe, and the "statistical" ones for collections of large numbers of constituents, be they atoms, stars or galaxies.

10.4 How Certain Are Our Conclusions?

After discussing the uncertainty which exists in the Universe we wonder how certain we can be about our cosmological conclusions. Questions of the certainty about scientific results affect all sciences, but this question is particularly accute in cosmology, for two reasons. First, because cosmology uses results from many other branches of science, which in turn are based on various assumptions, so we need to know what is certain and what is in doubt. Second, the conclusions of cosmology have very important implications. Many people would like to base their theories, and even their whole philosophy, on these conclusions.

We have to admit that the history of cosmology has disappointed us quite often. The most serious disappointments are related to the length-scale of the Universe and the question of whether the Universe is finite of infinite. The revisions of the distance scale over the past fifty years, which led to an almost continuous increase in the estimated dimensions of the Universe, are well known. The first distance scale for the distances of the galaxies was calculated by *Hubble*, around 1930. In 1950 all distances had to be doubled due to *Baade's* work. Later on, all extragalactic distances had to be multiplied by a factor of 5 because of *Sandage's* results, so that the distances we accept today for extragalactic objects are about 10 times those initially thought back in the 1930's.

As to whether the Universe is infinite or finite, before 1970 the idea of a finite pulsating Universe was widely accepted, then until recently it was thought that the Universe is infinite and continuously expanding, while today most people would accept that the actual Universe is so close to an exactly "flat" one that it is out of the question to state whether it is finite (with positive curvature) or infinite (with zero, or negative, curvature).

After all this, how can anyone trust the cosmological theories and conclusions? This is why some people, particularly non-specialists, view the subject with reservation.

The answer, however, is simple: "Our trust in cosmology is essentially the same as the trust in any other branch of science". Indeed, in all sciences we pass through periods of uncertainty and doubt, or even of mistakes, before we reach some definite results. This does not mean that science is untrustworthy. In spite of all these uncertainties, in spite of all the inaccuracies, and even in spite of all the mistakes, science progresses.

The idea some people have about science, that it is a series of revolutions, each one overturning previously held ideas, is wrong. On the contrary, the most important revolutions in science simply "correct" and complement the results of the past, without overturning previous scientific discoveries. The major change they bring about is that they focus the *interest* of the scientists on new points, where exactly the new research is being carried out. For example, the discovery of elementary particles did not abolish the concept of atoms, since almost all the matter in the Earth and the Universe is "still" made up of atoms. Scientific interest was simply transferred to the few but essential cases where the atoms and their nuclei change. Therefore, what the old physics was saying about the atoms is generally correct and the only change brought about by the new physics is that the deeper structure of the atoms has been discovered and studied. A similar process was followed in electromagnetism, relativity, astronomy and cosmology.

"Scientific discovery", says *Eddington,* "is like the fitting together of the pieces of a great jigsaw puzzle; a revolution in science does not mean that the pieces already arranged and interlocked have to be dispersed; it means that in fitting on fresh pieces we have had to revise our impression of what the puzzle-picture is going to be like. One day you ask the scientist how he is getting on; he replies "Finely. I have nearly finished this piece of blue sky". Another day you ask how the sky is progressing and are told, "I have added a lot more, but it was sea, not sky; there is a boat floating on top of it". Perhaps next time it will turn out to be a parasol upside down .... These revolutions of thought as to the final picture do not cause the scientist to lose faith in his handiwork, for he is aware that the completed portion is growing steadily".

In the same way every new discovery widens our understanding of the Universe without disproving our previous ideas (unless it is a case of an error) but it restricts or extends the area of their applicability.

The only mistake some people make is to consider their ideas absolute and to insist on applying them to cases where their application is entirely doubtful. If one, for example, insists that it is impossible to convert matter into energy because of *Lavoisier's* principle, it is a mistake because this principle has never been checked to sufficient accuracy to exclude the conversion of matter into energy. *Lavoisier's* principle holds generally, but ceases to hold exactly where the nuclear reactions start. The fact that certain unjustifiable generalisations were made of certain physical laws does not justify a general distrust of science. It only cautions us about any new generalisation we may wish to make in the future. It is, of course, understood that every theory, every measurement and every calculation has been adequately checked and that it is not the claim of one scientist or group of scientists alone.

Similar arguments hold in cosmology as well. The attitude "since I am not sure, I doubt everything" is not a scientific one, but one of mental laziness and inactivity of the brain. In general, doubt is beneficial to science, provided that any doubter takes the trouble to examine and check the validity of the results in question.

Another important point which must be stressed is the following: The degree of accuracy of the various results depends upon the kind of results and on the branch of science they come from. For example, although we know the electronic charge to an accuracy of 7 significant figures, the gravitational constant G is only known to 4 significant figures. This does not mean that electromagnetism is a more sophisticated science than gravity. We simply know the accuracy of our calculations in both cases.

The same holds for cosmology. We may measure the redshift of distant galaxies with an accuracy of 1%, but the age of the Universe has much greater uncertainty $(1 - 2 \times 10^{10}$ years). Indeed, Hubble's constant, which defines the distance scale of galaxies, and the age of the Universe, is estimated to be between 50 and 100 km/sec/Mpc, that is, there is an uncertainty there of a factor of two.

Then the question arises as to what sort of science one can do when the uncertainties are so large. The answer is again simple: cosmology will be affected very little if Hubble's constant is 50 or 100, because the revision of the value of H will simply change the scale of the various distances, but not the successive order of galaxies in distance from us, or the order in which the various phenomena occur during the evolution of the Universe. Thus, a change in Hubble's constant by a factor of 2 or even 10 does not change the sequence of cosmological phenomena (creation of elementary particles, creation of the first composite elements, creation of galaxies, stars and planetary systems).

Only when we want to compare the various independent estimates of the age of the Universe are we interested in finding Hubble's constant as accurately as possible. For example, we would be in trouble if we found a galaxy older than the age of the Universe[3]. The uncertainties in the various quantities are so great, however, that no such discrepancy appears in the data we have today.

---

[3] During the first estimates of the expansion of the Universe, it was thought that there are stars which are older than the Universe. This, however, has been corrected today.

11. Cosmological Problems

11.1 Inflation and Causality

One of the most important problems concerning the very early Universe is the "horizon problem". According to the standard Big Bang theory, the "radius of the Universe" (that is the scale factor R) during the early Universe was increasing with time as

$$R \propto t^{1/2}, \qquad (11.1)$$

where t is the age of the Universe. This implies that the expansion is fastest when $t \to 0$ when, in fact, the derivative of R tends to infinity. That is, the beginning of the expansion is an enormous explosion with infinite velocity.

On the other hand, the horizon of each particle has radius

$$R_* = c t. \qquad (11.2)$$

For each particle the "visible Universe" is a sphere with radius R_*. This is the region inside which it may exert its influence by emitting light signals or particles. This radius is zero at the beginning and increases with the speed of light.

In general, the radius of the visible Universe is greater than the scale factor R because it increases faster than $t^{1/2}$. During the very first stages of the life of the Universe, however, t was much less than $t^{1/2}$ and therefore the "radius of the visible Universe" for each particle was much less than the ordinary radius of the Universe, which is equal to the scale factor R. This means that the younger the Universe was, the smaller was the region influenced by each particle, with the ratio R_*/R approaching zero as t tended to zero. The implication is that the various parts of the Universe were independent of each other as $t \to 0$. We may wonder, therefore, how come the Universe has the observed isotropy and homogeneity since its various parts were uncorrelated at the beginning.

A solution to this difficulty is believed to be provided by the "inflationary" theory of the early Universe (Sect. 8.8). The inflationary scenario introduces a tremendous, inflationary expansion when the Universe was 10^{-35} sec old. In the period between 10^{-35} sec and 10^{-32} sec the expansion of the Universe was not given by (8.4) but by an exponential law like (6.6) (de Sitter Universe) with a very large cosmological constant λ. The resulting expansion was so large that a region smaller than a proton became larger than the whole of the observable Universe today. Therefore, the question why the various parts of the Universe are so similar is overcome by the fact that all these parts occupied a space much smaller than a proton when the age of the Universe was 10^{-35} sec.

But then the same question arises again. What happened before the inflationary period of the Universe? Even if the scale of the Universe was then 10^{50} times smaller than at present, the "horizon problem" comes back, and is even more difficult to solve than before. How did the various parts of this microscopic Universe communicate with each other before the inflation started? The *crucial* point is that the inflationary period started at a *finite* time, near the Grand Unification time 10^{-35} sec (or near the Planck time 10^{-43} sec) and not at time zero. Therefore, the very early expansion of the Universe (much before 10^{-35} sec) did follow the law (8.4) and the various parts of the very early Universe did not communicate with each other.

So, the basic causality problem related to the "horizon problem" (how can causally unrelated regions have the same properties) remains unanswered. This problem is one of the most profound unsolved problems of general physics. It does not only refer to the structure of the initial Universe (uniformity of initial conditions) but also to the structure of the physical laws themselves. The question is not only why the physical laws are the same everywhere in the Universe, but also why they are what they are and not different.

Several people have recently put forward the idea that the whole Universe originated as a result of a huge quantum fluctuation. Thus the creation of the Universe is not considered as a classical singularity in spacetime, but as a quantum effect. However, this theory is hardly a better "explanation" than the "original singularity" theory. The unsolved problem now is why spacetime had such properties as to allow a spontaneous quantum creation of the Universe.

11.2 Space and Time at the Beginning of the Expansion

When we construct models of the expanding Universe we are in danger of only solving some mathematical exercises which have nothing to do with reality. "Contact with reality" is necessary in the physical sciences and can be achieved in two ways: Firstly by using concepts with some physical meaning, and secondly by confirming theoretical conclusions with observations.

The "physical meaning" of a concept is specified by the way it is defined in physics. For example, when we talk about space and time we mean the measurements and observations which define these quantities. Lengths are defined by the ususal methods of physics and astronomy. The unit of length for small distances is the standard meter, or the wavelength of a specific spectral line. For astronomical distances we use trigonometrical methods, the comparison of apparent and absolute magnitudes etc. For extragalactic distances we apply photometric techniques: we assume the absolute luminosity of a galaxy or a star is fixed and by comparing it with its observed luminosity we calculate the distance. The clocks used to measure time are some physical or astronomical phenomena with periodicity, or more generally with *known* variability. Some of the commonest clocks are the motions of the planets, various atomic phenomena such as radiation at particular frequencies, the radioactive decay of certain elements or the changes of atoms and elementary particles with known half-life.

Having this in mind we can give physical meaning to the cosmological phenomena we observe. For example, when we say that the Universe expands, we mean that the usual distances between the galaxies change if measured with our usual units, such as the wavelength of a certain spectral line of the hydrogen atom. It is very important to use the correct unit, because if our unit expands as well, we shall not observe any change in the extragalactic distances. Therefore cosmic expansion exists because the meters we use are not subject to the same expansion. The same holds for time. We find the "age of the Universe" to be finite by comparing the cosmic expansion with certain atomic phenomena which are used for the definition of the unit of time. Thus we say that the Universe was expanding more rapidly in the past, and more specifically that its expansion rate increased indefinitely as time was tending to zero. We could not have concluded that, if the atomic phenomena were occuring at different rates during the early stages of the cosmic expansion. For example, time dilation in the special theory of relativity would be meaningless if the "moving" observer could not compare his time with that of observers *with respect to whom* he moves. An astronaut who travels to the distant stars with a velocity approaching that of the light (with respect to the Earth), does not feel that his life has been slowed down. He has the same biological cycles as the man on the Earth. He feels, for example, the need for 8 hours sleep every 24 hours, similar needs for food etc. It is only when the astronaut returns to the Earth that he sees that he is still young in comparison with his contemporaries on Earth, who will have aged or died.

But what is the meaning of time in cosmology? The problem is particularly relevant to the first stages of the evolution of the Universe when no atoms or composite nuclei were present; instead, there were only elementary particles changing continuously. Then on which cosmic phenomenon is the concept of time based? Similarly how could one measure distances and even define the notion of space, when the whole Universe we see today had the dimensions of a proton?

The answer to the above questions comes from theory. The relevant theories of nuclear physics on the reactions between elementary particles give us the half-life for the various decays. Therefore, the timescale during the first stages of the cosmic expansion is defined by the decay rates of elementary particles which were, in general, unstable. The degree of instability of those particles supplied the measure of time in the Universe. Similarly, the length scale was directly related to the intensity of the reactions between the particles which made up the Universe. Of course, no observer was present to measure time or space. But nor were there any observers when the Earth was formed or when life first appeared upon it. This does not mean that time did not exist "then". Time was simply specified by the competing rates of the various physical phenomena. The same happened during the first stages of the expansion of the Universe. As long as the theory concerning the decay of elementary particles is correct, we may apply our conclusions up to the point where we are met with some insuperable difficulty, even if we talk about times as early as 10^{-43} sec. The only condition is that our conclusions should not contradict current observational data.

However, the "beginning of the Universe" is indeed an insuperable difficulty. If the "initial" density was *infinite*, then the whole Universe was a mathematical singularity and the notions of space and time were meaningless. Similarly we lose the concept of space and time if we assume that the Universe started as a huge quantum fluctuation. For in this case, there was nothing "before" the Universe was created to undergo such a fluctuation. Thus, when we say that the Universe "started" at time $t = 0$ (i.e. at the beginning of the expansion) we mean that as we approach $t = 0$ the concepts of space and time become vague, until they are lost entirely. The well known sentence of *St. Augustine* expresses precisely that: "Non in tempore sed cum tempore finxit Deus mundum" (God created the world not in time, but with time). This is an idea essentially based upon *Plato*.

Occasionally scientists have used certain tricks to avoid the "origin of time" and its philosophical implications. The first is the change of timescale. We may devise "another time" t' which is the logarithm of the ordinary time:

$$t' = \log t. \tag{11.3}$$

This "other time" t' tends to minus infinity when the time t tends to zero. Therefore, although the Universe has a certain age in time t (say 20 billion years) it is still infinitely old in time t'. This trick was first proposed by *Milne* whose Eq. (7.3) is similar to the one we give above (11.3).

It is beyond doubt, however, that such a solution is entirely "artificial". It reminds us of Zeno's paradox about Achilles and the tortoise[1] which is based exactly on a change of timescale such that a finite time appears infinite.

Indeed, if the "creation" of the Universe is incomprehensible to the human mind, the difficulty does not go away with a mathematical trick like (11.3). The problem remains "why in the usual (or atomic) time t the age of the Universe is finite".

The other trick used to avoid the "beginning of the Universe" is to say that the Universe existed in some way before the Big Bang. That is, could it not be that the "initial explosion" was the result of a preceeding contraction of the Universe? The answer to this question is that whatever we say for the Universe before the initial singularity, we cannot rely on the known physics or on a mathematical extrapolation of the known physical laws. Essentially one can make *any* assumption about the "precreation" Universe. All assumptions, however, will be arbitrary as long as one accepts that the whole Universe passed through a singularity of infinite density.

This situation is similar to the following hypothetical case. If the Earth were to collide with the Sun, obviously it would vaporise. Now imagine astronomers from a distant planet trying to assess what the Earth was like before its destruction, and what civilisation it had developed, by observing the Sun after the collision. Whatever they assume will be arbitrary since nothing of that civilisation will have survived. Thus, it would be impossible for them to conclude either that the Earth was a dead planet, or that it had developed a high level of civilisation.

In the case of the Universe our ignorance is even more absolute. After all, we do observe the occasional collision of celestial bodies, but we never witness the creation of a Universe. Further, the hypothetical collision of the Earth with the Sun will destroy the structure of the Earth but not the chemical elements of which it is made. In the superdense conditions at the "beginning of the Universe", however, neither the chemical elements nor the elementary particles can survive.

Apart from the above arguments, another important consideration about the origin of the Universe refers to its "organisation". The evolution of the Universe is followed by a gradual decrease of its organisation. This

---

[1] Zeno's paradox runs as follows: Achilles runs to catch a tortoise. His speed u is, say, double that of the tortoise. If the tortoise's initial distance is a, Achilles needs time $t_1 = a/u$ to cover this distance. In the meantime, the tortoise has run on a distance $a/2$. Achilles requires time $t_1/2$ to cover this extra distance and so on. Zeno concluded that Achilles will *never* catch up with the tortoise because the tortoise will *always* be a little further ahead of the point Achilles will have reached. The answer to the paradox is that the infinite sum of the time intervals $t_1 + t_1/2 + t_1/4 + \cdots$ is finite (equal to $2t_1$). Only if we were using an artificial time t' with respect to which, for example, the successive time intervals were equal, would we find an infinite total time. In that case, however, both Achilles and the tortoise would be running with continuously decreasing speeds.

is known as the "second law of thermodynamics" or the law of "the increase of entropy" or "the increase of probability" (Sect. 9.1.1). The entropy increases as the Universe evolves, while it decreases if we go backwards in time. This means that the Universe is more "organised" as we go back in time. This organisation, is also called negentropy or anti-chance. The notion of negentropy is widely used in "information theory", which has so many applications to computer science. In the usual formulation of statistical laws, however, we usually avoid this concept. In fact we consider the evolution of a system in the positive direction of time, along which entropy increases and organisation decreases. Thus organisation or "anti-chance" is to be found in the "initial conditions" and not in the differential equations that describe the evolution of the system. But this separation is somewhat artificial and cannot be applied to the initial state of the Universe. In this case physical laws and initial conditions (or boundary conditions) are so intimately connected that we cannot separate them. The origin of time is characterised by the minimum value of the entropy, which could not be less than zero. Therefore, the origin of time is not a conventional origin like the beginning of the calendar year. It is a beginning when the "organisation" (or "negative entropy") of the Universe was a maximum, that is, it is not even conceivable to have had a preceding state. Eddington describes this situation as follows: "We have swept away anti-chance from the field of our current physical problems, but we have not got rid of it .... . Milliards of years into the past we find the sweepings piled up like a high wall, forming a boundary – a beginning of time – which we cannot climb over".

In more practical terms we may say the following: As we saw in Sect. 9.2 a little after the Big Bang the entropy of the Universe was 10^9 (approximately the ratio of photons to nucleons) while at the end (if the Universe is pulsating) it will be 10^{40}. That is, the entropy will increase by approximately 10^{40}. If we suppose that the "initial state" of the Universe was the "final state" of some previous existence, at the beginning of the previous existence the entropy would have been negative, which is impossible (since the ratio of the number of photons to nucleons cannot be negative). This confirms our view that the beginning of the Universe is not an arbitrary origin of time but something fundamental. There was no time before that beginning.

11.3 Is the Universe Cyclic?

The idea of a cyclic Universe appeared in various mythologies and has been supported by philosophers like *Nietzsche*. In modern times the theory of the pulsating Universe has been considered sometimes in the same framework. Namely, if the Universe is closed and evolves from the Big

Bang to the Big Crunch, then the end of our Universe is identified with a new beginning, the Big Crunch is the Big Bang for a new Universe.

We have already discussed the two basic difficulties of this picture (Sect. 9.2). We have seen that such a "new" Universe is not causally related to the old. Even more important is the fact that the entropy of the Universe undergoes a tremendous increase from its initial to its final state. The end of the Universe is in no way similar to its beginning.

In order to avoid these difficulties many efforts have been made to find models of the Universe that are free from singularities. In particular, some people have looked for models that have only closed time-like curves. Although such solutions of Einstein's field equations are highly improbable, they are not completely excluded. What would be the implication of such a model?

Usually we exclude a priori all models that allow closed time-like curves. The justification is that such models allow the violation of causality. We have already seen the absurdity of a situation in which a man returns to his own past and kills his alter ego. The absurdity arises from the fact that we have competing descriptions of the same present situation: "man alive" versus "man dead".

If all time-like curves are closed, however, everything repeats itself exactly in the same way, therefore there is only one description of the evolution of the Universe and not many contradictory ones. The situation is similar to that of a pendulum without friction that repeats the same motion for ever and ever.

Such a model of the Universe is possible but it does not look like *our* Universe. In fact, let us consider certain consequences of such a theory:

a) The Universe is exactly deterministic down to its most minute details. Not even the slightest indeterminacy is allowed. In fact, as we have seen (Sect. 10.3.4) even the quantum indeterminacy can produce very important deviations that are inconsistent with a cyclic world.

b) Despite its determinism, this system does not relate causes and effects. The condition of the Universe at a certain moment is both the cause and the effect of the condition of the Universe at another moment, past or future. There is no arrow of time from a cause to its effect.

c) There is no increase of entropy. In fact, the entropy of the Universe is exactly the same after each cycle.

d) As a consequence, all human actions are completely determined. This determinism also applies to the formulation of scientific theories, including the pulsating theory of the Universe. The author of the theory cannot do anything else but formulate this theory (in fact an infinite number of times, once per cycle) and there is no possibility of an independent check of its truth or falsehood.

But none of these phenomena is consistent with what we know of *our* Universe. Our Universe is basically causal but in many cases there is a distinction between cause and effect. The Universe has an entropy which increases. It has certain indeterminacies and it is not certain that even Einstein's theory of relativity is absolutely exact. Finally, human actions are partly free and one of the most important expressions of this freedom is science, with its continuing search for truth.

Therefore a cyclic Universe, although it may have some nice mathematical properties, cannot be considered as a valid model of our Universe.

11.4 Is There an "Ultimate" Theory of the Universe?

There are a number of people today who believe that we are close to the "end of physics". This belief is based on the success of the unified theories of the Universe (Sect. 8.5). These theories seem to explain many basic phenomena of physics and astronomy, like the symmetries between the quarks and leptons, the quantization of electric charge, the dominance of matter over antimatter, the homogeneity, isotropy and flatness of the Universe (through the inflation of the Universe), etc. If the basic prediction of the grand unified theories, namely the decay of the proton, is verified experimentally, then we will have to consider these theories as well established. Of course, one will need to select the most appropriate version of GUT, but this we may consider as a detail.

After the grand unification, the main problem of physics and astronomy is the supergrand unification, that will also include gravity. Although this goal still seems remote, many people work on it and have produced various theories that combine gravity with the other forces of nature. Such a theory is the "$N = 8$ supergravity" theory that is believed by *Hawking* to be the most successful theory of this kind. Another ambitious theory has recently been proposed by *Nanopoulos* (1984), and finally a very promising theory that considers the elementary particles as strings, rather than points, has been developed (1985). This last theory is the "superstring theory", which uses a spacetime of 10 dimensions, of which 4 form our usual spacetime, while the other dimensions are assumed to be "compactified". It is not possible to say yet what will be the ultimate fate of these ambitious theories.

But does an ultimate theory of the Universe exist? *Hawking*, who formulates this problem, considers three possibilities:

1. There exists a complete unified theory.
2. There are infinite levels of deeper and deeper theories.
3. There is no ultimate theory. Beyond a certain point observations cannot be described or predicted; they are just arbitrary.

Hawking believes that possibility (3) is eliminated by being incorporated into quantum mechanics. According to this point of view quantum mechanics is essentially a theory of what we do not know and we cannot predict.

On the other hand, *Hawking* believes that possibility (2) is unlikely, because there is a finite cut-off of the various physical scales at the Planck length and there does not seem to be any deeper level of smaller scale phenomena. This leaves us with possibility (1). This is why *Hawking* believes that the end of physics may be close, perhaps around the end of the century. The idea is that if a supergrand unified theory is proved to be correct, then the "goal of theoretical physics" will be satisfied and after that people will only have to work out the details.

But how can one be certain that he has found almost everything essential in physics and astronomy?

A similar belief that physics was approaching its end was widespread towards the end of the 19th century. The mechanistic theories of matter at that time seemed to explain almost everything. Only a few unexplained phenomena remained, like the exact nature of light (particles or waves). Many people expected to solve these problems soon and thus reach a complete understanding of nature. But then, at the end of the 19th century, in 1900 exactly, came Planck's theory and a little later, in 1905, the theory of relativity. These two theories destroyed the naive mechanistic views of the 19th century and started a new era full of revolutions and unexpected developments in physics.

Now, after one more century, we know incredibly more than at the end of the 19th century. Progress in physics has had an exponential expansion due to the dramatic increase of the number of physicists, equipment, and computers. In particular, the computers have considerably increased the possibilities in the development of theoretical physics. We can say that the human mind has acquired a new dimension of possibilities with their help.

In view of these developments, it is a little strange to claim again that we are approaching the end of physics. The historical evidence does not allow us such a pessimistic (or optimistic, for others) outlook. In fact, before every major new discovery it was not evident that something important was really around the corner. Only a few privileged people may have had this particular intuition that led them towards discoveries that were unexpected by the great majority of phyisicists.

Therefore, one cannot claim that there are no qualitatively new phenomena to be expected, which may reveal a deeper structure of the Universe. If we may venture to make our own guesses about the evolution of physics, we would like to mention two possible developments:

1) The most important recent advances in physics deal with certain basic symmetries. Such are the symmetries of quantum chromodynamics, the supersymmetries between hadrons and leptons, etc. However, it is possible that these symmetries are only approximate.

We have in mind the particular symmetries offered by an interesting class of dynamical systems that possess a number of exact integrals of motion, the integrable systems. The symmetries of such systems are sometimes hidden, but their consequences are always impressive. It is well known, however, that integrable systems are exceptional. In a certain sense they have a probability zero, while the probability of systems being "close" to integrable is not small. The behaviour of such systems is, in general, similar to that of an integrable system. But there are exceptional cases where the deviations become strong and new kinds of phenomena appear.

In the same way we expect that the symmetries found so far in elementary particles may be only approximate, although the approximations may be sufficient in most cases. If there are deviations from these symmetries one may need a deeper theory to describe them. This could mean new kinds of elementary particles, and perhaps also new forces, beyond the four basic forces which we currently accept. Such possibilities remain open and no one can exclude them a priori.

2) Another area of new developments refers to complex systems of many bodies that may range from complex particles and nuclei, to chemical compounds and living organisms. The possible states of many body systems are so numerous that they cannot be explored systematically even with the help of the most powerful computers. It is unbelievable how many possibilities appear in deceptively simple looking systems of 2 or 3 degrees of freedom. The possibilities offered by systems of many degrees of freedom are really astonishing. One can always find new types of systems with unexpected new properties which cannot be predicted a priori. This is particularly true with regard to complex organic molecules and living organisms. One cannot say that the properties of these systems are derivable from the properties of their simple components, because all these effects are highly nonlinear.

The complex entities found in chemistry and biology (e.g. molecules, cells and organisms) may be building blocks for even more complex structures. Thus molecules are entities of a higher level than atoms, cells are of higher level than molecules, organisms are of a higher level than cells, and intelligence is of an even higher level, even though each level is composed of elements taken from the previous levels.

We consider the exploration of these highly nonlinear systems an extremely rich area of physics (including chemistry and biology), although one may not need new types of forces in order to account for them.

Thus, among the three possibilities concerning an ultimate law of physics we think that the most probable is the second one, of infinite levels of knowledge.

As regards the third possibility it is doubtful if one can derive knowledge out of complete ignorance. Statistical laws are not based on complete ignorance, but on some knowledge of the behaviour of the elements of an ensemble. This is how we derive the nice Gaussian distributions that are the basis of our statistics. For example, one could never derive the laws governing the kinetic theory of gases if every molecule behaved in a quite arbitrary way.

We hope to understand better the quantum phenomena, and any other statistical phenomena, with more knowledge about them, and not by an appeal to our ignorance.

In conclusion, we believe that we are as far from an end of physics as ever. We only hope to be able to say that we understand our Universe better than previous generations of physicists.

11.5 Causality and Teleology

When we ask "why" a certain event has happened we may justify it as a result of some cause, or attribute it to a conscious and purposeful action. For example we may ask: Why are there so many people in the museum today? The answer may be "because there is no entrance fee today" (cause) or "because they want to see a new exhibit" (purpose). The cause and purpose are very often related in events concerning people. Also, causes which concern people may be different from those concerning material objects. A characteristic example comes from *Socrates* in Phaedo. Socrates is in prison, having been condemned to death by his countrymen. Before his death he calmly discusses philosophical questions with his disciples. He criticises the natural philosophers who stress the most trivial aspects of his actions and neglect the moral causes behind them: "Trying to account for the cause of my actions, they say first that I am sitting here because my body consists of bones and nerves which as they stretch and contract make my limbs to bend in the positions they have at this moment, while they fail to mention the true cause. Because these bones and nerves would have been in Boeotia ages ago (that is if he had accepted to bribe the guard and escape, as was suggested to him) if I did not believe that it is fairer and kinder to accept the decision of my country instead of leaving".[2]

The natural philosophers are, of course, right but their explanation brushes aside the most important "moral" aspect of Socrates' action, his moral ideas which make him remain in prison awaiting death instead of fleeing to freedom.

Our attitude to physical phenomena, however, is exactly the opposite. The only explanations which we accept in such cases are the causal ones. For example, the geometric propagation of light may be explained in two

---

[2] Plato, *Phaedo*, Sect. 47.

ways: The "causal" explanation is based upon the concept of light rays emitted from a source and propagating according to Maxwell's equations. On the other hand, the "teleological" explanation is based on the principle of least time. Light travels from point A to point B by the shortest route. This path is not always a straight line. When light is refracted, for example, it follows a non-straight path, since this is the fastest route by which the light may travel. Should one therefore say that the light knows where it wants to go and chooses the fastest route? Of course not. That is why the principle of least time is considered only as a mathematical game and not as the basic explanation of the propagation of light.

There are other phenomena, however, that refer to life, which show a definite purposefulness, for which we cannot find a causal explanation. For example, the instincts of animals and the evolution of life itself are wonderful examples of this purposefulness. These phenomena are not understood in terms of causality. Many people *hope* that one day all these phenomena will be explained causally, but this is far from certain.

In any case, the causal explanations always suffer from a fundamental imperfection. They are never complete. Any answer we may give to a question "why ..." satisfies us only temporarily because another question "why ..." more difficult and demanding soon presents itself.

A child may ask: "Why is it light during the day, and dark at night?" Of course, the answer is not that "because that is how we observe it to be" but some explanation is offered: "Because during the day we have the Sun which produces light". The child, however, continues: "Why does the Sun produce light?". The answer is now more difficult. We usually recommend the child to study first and then we tell him: "Because nuclear reactions which occur in the Sun produce energy". The answer may be satisfactory if backed up by enough nuclear physics and mathematical formulae, but only for a short while. The questions will come back again, deeper and more persistent: "Why do hydrogen, helium and the other elements have the properties they do? For example, the nuclei of these elements have potential barriers which, however, may be penetrated by particles approaching them (tunnel effect), and produce nuclear reactions. Then the extra mass is converted into energy. Why is this so?". At this point one mobilises all his knowledge of quantum mechanics, relativity, high energy physics and even the latest theories of unified fields in an attempt to answer all these questions. But the questions remain: "Why do quantum mechanics, relativity or unified field theories have the properties they do?". A moment will come when the father will answer with indignation (or alternatively in a "philosophical" way): "Because that is how it is". But then why didn't he give this answer right at the beginning, to the first question asked by the child: "because that is how it is"? At this point the philosophical intention of the father becomes clear. He has been trying to make the child resign from the endless chain of questions "why ...". He believes that he might persuade the child that he is demanding too much, and that it is enough to be able, or soon to be able (with our unified field theories), to unite in one theory all the forces of nature, from the strong force to the gravitational force.

Nothing assures us, however, that tomorrow we are not going to discover new phenomena which will demand an even deeper theory, or that we are not going to find exceptions to our theories, or even new aspects of them which will have new implications.

It is not at all certain that we have discovered the perfect theory which governs the Universe and its laws. Even if we find such a theory the question will always remain: "why is the Universe and its laws this way and not different?"

At this point one may put forward a different question: "What would have happened if the laws of the Universe were different?". The most interesting exercise is not to find out what would happen to any model universe, but to a universe only slightly different from our own.

The most important result of such research is that if the laws which govern the Universe were different, life and man could not exist in it.

Such an "Anthropic Principle" has been proposed by *Carter* (1974), but similar observations have been made by *Dicke* (1961), *Hawking* (1974), *Wheeler* (1974), *Rees* and *Carr* (1979) and others. *Carter*'s observation stems from the strange coincidences of large numbers of the order of 10^{40} which we find in the Universe (Sect. 7.5). These coincidences led *Dirac* to propose his revolutionary ideas about the variability of the gravitational constant G and the creation of new matter in the Universe, which are in contradiction with the general theory of relativity. But *Carter* gives a quite different interpretation of these coincidences. He claims that it is *because* of the coincidence of these large numbers that there is intelligent life in the Universe. This thought is based upon the suggestion that the age of the Universe is not arbitrary but of the same order of magnitude as the age of the stars of our Galaxy, which is estimated to be 10^{40} in the units mentioned in Sect. 7.5. Indeed, if the Universe were much younger than it really is, no stars would have been formed yet and if it were much older the stars would have died. In either of the above cases no life would exist in the Universe. That is, the age of the Universe is directly related to the existence of life. *Carter* uses similar reasoning to show that if the number of particles in the Universe (which is related to the density of the Universe) were very different from 10^{78}, life would not exist.

Papagiannis (1978) presents a more detailed account of the characteristics of the physical laws that favour life. The appearance of intelligent life in the Universe is based on the following properties of the laws of physics: (A) The property of atoms to combine into complicated molecules. (B) The existence of sources of energy which last several billions of years so that higher forms of life may have time to evolve. (C) The appropriate form of the energy created such as to favour life instead of destroying it (e.g. the temperatures involved are neither too high nor too low).

A) The formation of complicated compounds of matter is based on the existence of the strong, weak and electromagnetic forces. If nature had wanted to make dead stars, one type of neutral particle and one kind of force, gravity, would have sufficed. The formation of composite atoms, however, depends upon the strong and weak nuclear forces. The formation of complicated molecules, which are necessary for life, depends upon the electromagnetic forces. The significance of these forces is particularly clear in the case of carbon, which is the basic element of life. Carbon has the amazing property to form an endless number of compounds, up to the complex DNA molecules which are the foundation of life. Apart from carbon only silicon has similar properties, but to a lesser extent. This is why life on Earth is based on carbon chemistry and not silicon chemistry, even though silicon is ten times more abundant. The appearance of carbon is due to some peculiarity of the nuclear reactions in stellar interiors. As we explained in Sect. 6.7 elements heavier than $_2He^4$ could not have been formed during the Big Bang. Carbon can only be produced inside stars where at temperatures of 10^8 K, three helium nuclei combine to form one carbon nucleus. The fusion of two helium nuclei produces a metastable isotope of beryllium, $_4Be^8$, which then combines with a third helium nucleus to form carbon. The word metastable means that the nucleus is not stable, but it is not entirely unstable either. In isolation $_4Be^8$ would decay in about 2×10^{-16} seconds. In this time, however, a third helium nucleus is able to join it and thus carbon is produced. If $_4Be^8$ were entirely unstable, carbon could not have been formed and life could not exist. The length of time which $_4Be^8$ survives depends upon the intensity of nuclear reactions. Therefore, the strength of the nuclear forces is relevant to the existence of life in the Universe.

B) The existence of long-lived sources of energy (the stars) depends upon the slow conversion of hydrogen into helium. Hydrogen is still the most abundant element in the Universe because only some of it was converted into helium during cosmic nucleosynthesis, within the first three minutes of the cosmic lifetime. This is due to the fact that the nuclear forces between protons are just strong enough to allow some nuclear reactions but not too strong to make all protons combine in deuterium and helium nuclei.

If almost all the matter in the Universe were helium, the ordinary stars like the Sun would not exist, since they produce their energy from "burning" hydrogen (i.e. converting hydrogen into helium). If only helium existed we would only have stars which produce energy from burning helium. Such stars are, however, very short lived. In their short lifetime, of a few dozen million years, it would not have been possible for life to develop.

Further, the production of energy by the Sun is based upon the properties of the nuclear force. The first stage of producing helium from hydrogen is the production of deuterium. In the interior of a star, deuterium is produced by the conversion of a proton into a neutron by the reaction $p \rightarrow n + e^+ + v$, and hence $p + p \rightarrow D + e^+ + v$. This reaction is governed by the weak nuclear force, and is extremely slow. If this reaction were as fast as other nuclear reactions, the Sun would have exploded in a very short time, like a huge hydrogen bomb. Another factor which ensures the longevity of the Sun is the opacity of its outer layers. The electromagnetic forces which dominate the particles there ensure that the radiation coming from the nucleus is absorbed and thus high temperatures are maintained in the interior. If it were not for this absorption, the Sun would have been extinguished within a day.

Finally, if the constant of gravity, G, were different, the ages of the stars would be entirely different. Indeed, the luminosity of a star is proportional to the seventh power of G. That is, if G were two times its actual value, the stars would be 128 times brighter, and therefore their lives would be 128 times shorter. The evolution of life on Earth, however, took 3 billion years. A reduction in the ages of the stars by a factor of 128 would not have left enough time for that evolution to take place. If the value of G were smaller than it is, the formation of stars and planets would be much more difficult or even impossible.

C) Concerning the *form* of energy reaching the Earth, we notice that if the energy of photons were too high, implying a higher temperature on the Earth, then the organic compounds would have been destroyed, or if it were too low, everything would have been frozen and the organic molecules would not have the necessary mobility for the preservation of life. The ratio of the electromagnetic to gravitational forces must have a certain value for the photons to carry the right amount of energy. This ratio is

$$\frac{e^2}{G m_p m_e} \simeq 10^{40}, \tag{11.4}$$

where m_p and m_e are the masses of the proton and the electron respectively.

One might expect that a change of this huge number by a few orders of magnitude would not change matters, but this is not so. It seems that there is a sensitive balance between the various natural forces which cannot change appreciably without catastrophic consequences for life. We have already seen that a change in the value of G by only a factor of two would have made life highly improbable. The same holds for the rest of the constants in (11.4). For example, an increase in the electron's mass by a factor of 10 would have made the energies required for the various organic reactions of life ten times greater, and thus much more difficult to attain. On the other hand, the reduction of the electron's mass by a factor of ten would have made the organic molecules very unstable, being immediately destroyed by the solar radiation. Similar irregularities would be observed if the proton mass were different. In this case, the nuclear reactions in the solar interior would also be affected.

Finally, a change in the electronic charge *e* would change not only the stability of the organic chemical reactions, but also the rates of the nuclear reactions in the stellar interiors as well.

The above remarks show that it is not possible for the basic forces of nature (or the constants upon which they depend) to change appreciably, without restricting or even hindering altogether the development of life.

Similar thoughts have recently been expressed by *Rees* (1981). He observes that if the nuclear forces were slightly stronger, protons would have formed biprotons and therefore hydrogen, the basis of all other forms of matter, would not exist. On the other hand, if the nuclear forces were a little weaker no other elements apart from hydrogen would exist. The results would have been equally disastrous for life if gravity or electromagnetic forces were weaker. Thus, he concludes with the following version of *Carter's* anthropic principle (paraphrasing *Descartes* who said: "cogito ergo sum", I think therefore I am): "cogito ergo mundus talis est" (I think, that is why the Universe is how it is). As *Rees* observes, the fact that life was the outcome of apparently insignificant details in the structure of the early Universe can be characterised as either pure luck or divine providence.

This subject requires more study. In any case, many indications point towards a purposeful Universe. Examples of purpose concerning life on Earth have been noted many times in the past, but we shall not mention them here. The new element which we describe here is that "purposefulness" extends to the whole Universe. We do not refer simply to the favourable conditions for the development of life on Earth, but to the *prerequisites* for the existence of planets like the Earth, stars like the Sun to illuminate them and chemical compounds appropriate for the development of life.

The existence of purpose in the Universe is admittedly a sore point for many people. It bears witness to a plan of a conscious mind behind the Universe. For this reason several people doubt that there is any purpose in nature and in the Universe. There are two main objections to the existence of a plan behind the Universe:

1) There are cases where events are only apparently purposeful. The example of light which follows the shortest path has already been mentioned. The same phenomenon can be better explained by causality. Thus many people hope that at some time all phenomena will be explained similarly. But how can anyone be certain that all phenomena can be explained by causality? Many phenomena *cannot be explained without purposefulness*. For example, the archeologist who finds some carved stones may wonder whether they have been made by nature or by primitive people. When, however, he finds traces of fire and left over food he

will have every reason to believe that these tools had been made for a purpose. His belief will be reinforced when more complicated tools are found which cannot have been made by chance. Thus the question of the existence of a purpose in the Universe cannot be set aside by the claim that one day *it will* be shown that purpose does not exist.

2) Some people have expressed the point of view that the Universe *happened* to be like it is and *that is why* there is life and intelligent beings. If it were different, life would not exist and we would not be here to discuss it. Some people talk about an infinite set of universes and some even hint that the other universes "exist" (maybe in some other as yet unknown dimensions of space). The fact is that we are certain of the existence of only one Universe, our own. What can we say about the other hypothetical universes since we do not even have the slightest clue for their existence? How meaningful is it to talk about them?

By the same logic one may assume that all phenomena in the Universe are the result of chance and that there are no physical laws but only random coincidences. Such an extreme point of view, however, abolishes any concept of science.

Of course, there is a third attitude towards a purposeful Universe, one of reservation. There are many people who doubt that nature and man have some purpose. That is why they do not want to discuss "dangerous" matters such as this. But doubt in science is the motivation for research. Things go wrong when doubt becomes an opportunity for boasting and self-confidence. Socrates said: "ἓν οἶδα ὅτι οὐδέν οἶδα" (Only one thing do I now, that I know nothing). He did not say: "I know nothing, therefore I am a great philosopher". Exactly the opposite, he was searching for the truth all his life and he sacrificed his life for it.

Indeed, the basic problems of man are the meaning of life and death. Such problems cannot be solved either with unjustified optimism or with an arrogant approach. The inadequacy of the naive agnosticism becomes clear in matters like this. Such problems give us a glimpse of new aspects of reality that are not less important than the laws of physics.

11.6 Cosmology and Metaphysics

When one exhausts the chain of questions "why ..." concerning the Universe, he arrives at questions which remain essentially unanswered. *Rees* mentions a series of such questions in one of his recent articles (1981): "What existed before the initial explosion? Why is the Universe inhomogeneous on small scales and yet homogeneous on large scales? Why has space three dimensions and time only one? Why are the constants of nature so universal? And most of all why is the Universe as a whole so symmetric and simple, a necessary requirement for the remarkable progress made in cosmology?"

Maybe some of these questions will be answered one day. Essentially, however, the problems will be transferred to other more intractable ones. As *Einstein* said: "The most incomprehensible thing about the Universe is that the Universe is comprehensible.". Indeed, nothing could have guaranteed in advance that the Universe would be comprehensible to the human mind. Nothing tells us that the Universe ought to have the amazing unity we observe, or that it had to be such that it could be described by physical laws which hold everywhere, or that the physical laws had to be such that they could be described by mathematical formulae, and so on.

Many of the questions that the cosmologists end up to are of a metaphysical nature. This causes an embarrassment to some of them. For many people anything metaphysical is taboo. *Monod*'s statement is characteristic: "Any mixture of knowledge and values is illegal, forbidden". The irony of the matter is that *Monod*'s statement is based on the belief that knowledge is the supreme value. But who gave us the right to make such a claim? Isn't it an arbitrary and unfounded principle to put our brain as an undoubted supreme value, as an absolute and unquestionable authority?

There is something else, more basic. Is it ever possible to distinguish absolutely physics from metaphysics, and knowledge from values? As *Planck* said: "there is no physics without some metaphysics". Indeed the most basic questions of physics, like what is the cause of the whole Universe, or why physical laws are what they are, which we discussed in the previous section, can be characterised as metaphysical questions. This does not mean, however, that they are less important than the "physical" questions, like why does the Sun illuminate the Earth, or why does this or that nuclear reaction take place.

Metaphysics deals with the structure and foundation of physics, in the same way as "metamathematics" examines the structure and foundation of mathematics. Therefore, when one asks such basic questions he is not introducing theology into physics. In fact, there exists what is called "negative metaphysics", which denies any concept of God. Moreover, there are scientists who discuss metaphysics without calling it as such, either intentionally, or because they do not believe that negative metaphysics is a part of metaphysics. A characteristic example is the strong support given by some people to the theory of continuous creation, as a "scientific" theory, as opposed to the Big Bang theory which they see as more "theological". As many people have observed, however, the continuous creation is no less creation than the creation of the whole Universe, and therefore it does not cease to be a metaphysical notion similar to the concept of an initial creation.

Nowadays, of course, the theory of continuous creation has been largely abandoned, and discussions of this sort appear naive. On the other hand the fact that most astronomers and physicists accept the Big Bang theory does not mean to say that they all see that theory with the same metaphysical implications. One may find a great variety of points of view, from *Weinberg* (Nobel prize 1979) who finds the Universe to be pointless, although he accepts the Big Bang theory, to *Penzias* (Nobel prize 1978) who sees the current development of physics and cosmology as a verification of the Bible in its entirety.

In general, the scientists who deal with the Universe can be divided into two categories according to their attitude towards the Universe.

On the one hand those who see the wonders of the Universe either in a detached way, as one might watch a movie on television, or as a means of stressing their own importance. Every new discovery increases their confidence in human ingenuity, either that of their own minds or that of mankind in general. This attitude, which tends to turn man into a "god", has been called "ὕβρις" (insolence) by ancient greek philosophers.

On the other hand, there are those scientists that stand in front of the Universe with wonder and awe. Every new discovery they make shows them their limitations. They realise that Nature and the Universe are incomparably beyond man with his imperfections and limited capabilities. This does not lead them to desperation, but into a deeper certainty that the Universe, which is far above man's present knowledge and understanding, not only is not hostile to man but has a deep affinity with the most high qualities of man. It is this affinity that answers Einstein's basic question "why the Universe as a whole is comprehensible to man"; because both the Universe and man have the same origin, the same descent.

References

The bibliography below includes books which, according to our opinion, are useful for further study. From the older books we list only the best which include important and original ideas. Books published after 1984 are in general not included.

1. Abell, G., Chincarini, G. (eds.): *Early Evolution of the Universe and Its Present Structure* (Reidel, Dordrecht, 1983). Proceedings of the 104th Symposium of the International Astronomical Union in Crete. It includes the most recent developments in the major areas of cosmology.
2. *Astrophysics and Elementary Particles, Common Problems* (Accad. Naz. Lincei, Roma 1980). Proceedings of a meeting with interesting presentations by Glashow, Steigman, Sciama, Ellis, Carr etc.
3. Audouze, J., Balian, R., Schramm, D.N.: *Physical Cosmology* (North Holland, Amsterdam 1980). It includes the courses of the summer school "Les Houches" of 1979. Book of high level.
4. Audouze, J., Van Tran Thunh, J. (eds.): *Formation and Evolution of Galaxies and Large Structures in the Universe* (Reidel, Dordrecht 1984). It includes 35 papers on the early Universe and the structure of the Universe today.
5. Barnes, A.C., Clayton, D.D., Schramm, D.N. (eds.): *Essays in Nuclear Astrophysics* (Cambridge University Press, Cambridge 1982)
6. Bath, G. (ed.): *The State of the Universe* (Oxford University Press, Oxford 1980). A series of lectures on the origin of the Universe and the chemical elements, black holes etc. by Sciama, Rees, Tayler, Blackwell, Pounds, Penrose, Hunt and F.G. Smith. Very interesting.
7. Berry, M.: *Principles of Cosmology and Gravitation* (Cambridge University Press, Cambridge 1976). Book for the general public.
8. Bondi, H.: *Cosmology* (Cambridge University Press, Cambridge 1951). A well written book containing the ideas of one of the founders of the theory of continuous creation.
9. Bonnor, W.B., Islam, J.N., MacCallum, M.A.H.: *Classical General Relativity* (Cambridge University Press, Cambridge 1984). It contains a number of cosmological papers.
10. Bowler, M.G.: *Gravitation and Relativity* (Pergamon, Oxford 1976). Mathematical treatment, not particularly difficult.
11. Brück, H.A., Coyne, G.V., Longair, M.S. (eds.): *Astrophysical Cosmology* (Pontificia Academia Scientiarum, Citta del Vaticano 1982). It contains several very interesting articles by Hawking, Zeldovich, Weinberg etc.
12. Chandrasekhar, S.: *The Mathematical Theory of Black Holes* (Clarendon, Oxford 1983). Mathematical presentation. The best book in this field.
13. *Cosmology + 1, Reading from Scientific American* (Freeman and Co., San Francisco 1977). A collection of articles published at various times by the magazine Scientific American. All articles are very interesting.
14. Davies, P.: *The Accidental Universe* (Cambridge University Press, Cambridge 1982). An advanced book for the general public.
15. Deruelle, N., Prian, T. (eds.): *Gravitational Radiation* (North Holland, Amsterdam 1983)
16. Dewitt-Morette, G.: *Gravitational Radiation and Gravitational Collapse* (Reidel, Dordrecht 1974). Proceedings of the 64th symposium of the International Astronomical Union. Quite specialised.
17. Disney, M.: *The Hidden Universe* (Dent, London 1984)
18. Eddington, A.: *The Nature of the Physical World* (Cambridge University Press, Cambridge 1928). A very good presentation of the philosophical consequences of the theories of physics, the general theory of relativity included.

19. Eddington, A.: *New Pathways in Science* (Cambridge University Press, Cambridge 1934). A sequel to the previous book.
20. Ehlers, J., Perry, J.J., Walker, M. (eds.): *Ninth Texas Symposium on Relativistic Astrophysics* (N.Y. Acad. Sci., New York 1980)
21. Einstein, A.: *The Meaning of Relativity*, 4th ed. (Methuen, London 1950). It includes, in a simplified form, Einstein's ideas on the special and general theories of relativity, as well as his unified theory.
22. *Exploration of the Universe, ESO Symposium* (European Southern Observatory, Garching 1982). It contains some articles with particular cosmological interest.
23. Field, G.B., Arp, H., Bahcall, J.N.: *The Redshift Controversy* (Benjamin, Reading 1973) After an introduction by Field, Arp's and Bahcall's ideas are discussed and copies of relevant papers are reproduced.
24. Fiorini, E. (ed.): *Neutrino Physics and Astrophysics* (Plenum, New York 1982)
25. Gamow, G.: *The Creation of the Universe* (The Viking Press, New York 1952). An interesting book for the general public. It contains many original cosmological ideas by Gamow.
26. Gerbal, D., Mazure, A. (eds.): *Clustering in the Universe* (Editions Frontieres, Gif sur Yvette 1983)
27. Gibbons, W.G., Hawking, S.W., Silkos, T.S. (eds.): *The Very Early Universe* (Cambridge University Press, Cambridge 1983). Very interesting.
28. Harrison, E.R.: *Cosmology* (Cambridge University Press, Cambridge 1981). A very interesting book without difficult mathematics.
29. Hawking, S.W., Israel, W. (eds.): *General Relativity* (Cambridge University Press, Cambridge 1979). An important series of articles on relativity and cosmology by Hawking, Israel, Chandrasekhar, Will, Braginsky, Geroch, Carter, Thorne, Dicke, Peebles, Zeldovich, Weinberg, Penrose etc. Very high level.
30. Hawking, S.W.: *Mindsteps to the Cosmos* (Harper and Row, New York 1983)
31. Heidmann, J.: *Introduction à la cosmologie* (Presses Univsersitaires de France, Paris 1973). [English transl.: *Relativistc Cosmology* (Springer, Heidelberg 1980)]. An introductory book on cosmology of general interest.
32. Held, A. (ed.): *General Relativity and Gravitation* (Plenum, New York 1980). Two volumes. A series of articles on Einstein's centenary.
33. Hodge, P.W. (ed.): *The Universe of Galaxies: Readings from Scientific American* (Freeman, New York 1984)
34. Hoyle, F.: *Astronomy and Cosmology. A Modern Course* (Freeman, San Francisco 1975). A book for the general public.
35. Israel, W. (ed.): *Relativity, Astrophysics and Cosmology* (Reidel, Dordrecht 1973). It includes articles by Ehlers, Brill, Hartle, Sachs etc.
36. Jones, B.J.T., Jones, J.E. (eds.): *The Origin and Evolution of Galaxies* (Reidel, Dordrecht 1983). Proceedings of a conference on the formation and evolution of galaxies.
37. Kaufmann, W.J. III: *Relativity and Cosmology* (Harper and Row, New York 1972). A short descriptive book without any mathematics.
38. Kaufmann, W.J. III: *The Cosmic Frontiers of General Relativity* (Little, Brown and Co., Boston 1977). A detailed discussion of the subject of black holes for the general public.
39. Landau, L.D., Lifshitz, E.M.: *The Classical Theory of Fields* (Pergamon, Oxford 1962). A very good mathematical book of high level. It includes the special and general theories of relativity.
40. Landsberg, P.T., Evans, D.A.: *Mathematical Cosmology. An Introduction* (Clarendon, Oxford 1977). A mathematical presentation but without particular difficulties.
41. Layzer, D. (ed.): *Constructing the Universe* (Scientific American Library, New York 1984)
42. Levy, M., Deser, S.: *Gravitation* (Plenum, New York 1979)
43. Longair, M.S. (ed.): *Confrontation of Cosmological Theories with Observational Data* (Reidel, Dordrecht 1974). Proceedings of the 63rd symposium of the International Astronomical Union. It contains many interesting articles.

44. Lynden-Bell, D. (ed.): *The Big Bang and Element Creation* (The Royal Society, London 1983). Proceedings of a conference on the origin of the Universe and the formation of the elements, held in 1982. It includes the advances in cosmology up to that time.

45. Mavrides, S.: *L'Univers relativiste* (Masson et Cie., Paris 1973). A well written book combining observations and a mathematical treatment.

46. McCrea, W.H., Rees, M.J. (eds.): *The Constants of Physics* (The Royal Society, London 1983). It contains several interesting articles by McCrea, Ellis, Rees, Press, Barrow, Carter etc.

47. Meurers, J.: *Kosmologie heute* (Wissenschaftliche Buchgesellschaft, Darmstadt 1984) (in German). A discussion of the cosmological ideas with little mathematics.

48. Misner, C.W., Thorne, K.S., Wheeler, J.A.: *Gravitation* (Freeman, San Francisco 1973). A fundamental book of 1280 pages, with a wide mathematical analysis of the various subjects.

49. Narlikar, J.V.: *General Relativity and Cosmology* (Macmillan, London 1979). An introductory book with a mathematical presentation.

50. Narlikar, J.V., Kembhavi, A.K.: *Non-Standard Cosmologies* (Gordon and Breach, New York 1980). It includes a detailed discussion of the various cosmological theories with emphasis on theories other than general relativity.

51. Nelkowski, H. et al. (eds.): *Einstein Symposion Berlin*, Lect. Notes Phys., Vol. 100 (Springer, Berlin, Heidelberg, New York 1979). Proceedings of a symposion held on Einstein's centenary. It includes interesting articles by Ehlers, Penrose, Iliopoulos etc. but several articles are in German. About one third of the articles are on philosophical or historical matters.

52. Novello, M., Salim, J.M.: *Non-Equilibrium Relativistic Cosmology*, Fundamentals of Cosmic Physics 8, No 3, 1983

53. Novikov, I.: *Evolution of the Universe* (Cambridge University Press, Cambridge 1982) [transl. from Russian]. An advanced book for the general public.

54. Omnès, R.: *L'Univers et ses métamorphoses* (Hermann, Paris 1973). A book without mathematics aimed for the general public.

55. Papagiannis, M.D. (ed.): *Strategies for the Search for Life in the Universe* (Reidel, Dordrecht 1980). Proceedings of a special meeting organised by the International Astronomical Union on the subject of life in the Universe.

56. Papapetrou, A.: *Lectures on General Relativity* (Reidel, Dordrecht 1974). A course of postgraduate level.

57. Peebles, P.J.E.: *Physical Cosmology* (Princeton University Press, Princeton 1972). An advanced study with emphasis on the physical phenomena in cosmology.

58. Peebles, P.J.E.: *The Large Scale Structure of the Universe* (Princeton University Press, Princeton 1980). An updated book of high level.

59. Raine, D.J.: *The Isotropic Universe* (Hilger, Bristol 1981). An introductory book, rich in cosmological subjects.

60. Raychaudhuri, A.K.: *Theoretical Cosmology* (Clarendon, Oxford 1979). A relatively simple book without advanced mathematics.

61. Rindler, W.: *Essential Relativity. Special, General and Cosmological*, Texts Mon. Phys. (Springer, Berlin, Heidelberg, New York 1977). An important systematic book with mathematical presentation.

62. Robertson, H.P., Noonan, T.W.: *Relativity and Cosmology* (Saunders Co., Philadelphia 1968). An interesting systematic book with a mathematical presentation.

63. Ronan, C.A.: *Das Kosmosbuch des Weltalls* (Kosmos, Stuttgart 1983)

64. Rowan-Robinson, M.: *Cosmology*, 2nd ed. (Oxford University Press, Oxford 1981). A very good introductory book with little mathematics.

65. Ryan, M.P. Jr., Shepley, L.C.: *Homogeneous Relativistic Cosmologies* (Princeton University Press, Princeton 1975). A mathematically advanced book.

66. Sciama, D.W.: *Modern Cosmology* (Cambridge University Press, Cambridge 1971). A well written book without mathematics.

67. Sexl, R.U., Urbantke, H.K.: *Gravitation und Kosmologie* (Bibliographisches Institut, Zurich 1975). Very good mathematical presentation of the subject (in German).
68. Shapiro, L.St., Teukolsky, A.S.: *Black Holes, White Dwarfs and Neutron Stars. The Compact Objects* (Wiley Interscience, New York 1983)
69. Stephani, H.: *General Relativity* (Cambridge University Press, Cambridge 1982)
70. Straumann, N.: *General Relativity and Relativistic Astrophysics* (Springer, Berlin, Heidelberg, New York 1984)
71. Terzian, Y., Bilson, E.M. (eds.): *Cosmology and Astrophysics. Essays in Honour of Thomas Gold* (Cornell University Press, Ithaca 1982)
72. Van der Merwe, A. (ed.): *Old and New Questions in Physics, Cosmology, Philosophy and Theoretical Biology* (Plenum, New York 1983)
73. Wagoner, R.V., Goldsmith, D.W.: *Cosmic Horizons* (Freeman and Co., San Francisco 1982)
74. Wald, R.M.: *Space, Time and Gravity: The Theory of the Big Bang and Black Holes* (University of Chicago Press, Chicago 1977)
75. Weinberg, S.: *Gravitation and Cosmology* (Wiley, New York 1971). Very good book with advanced mathematical presentation.
76. Weinberg, S.: *The First Three Minutes* (Basic Books, New York 1977). Very interesting book on the origin of the Universe, without any mathematics.
77. Wesson, P.S.: *Gravity, Particles and Astrophysics* (Reidel, Dordrecht 1980). It discusses the various cosmological theories in comparison with observations, without any mathematics.
78. Wolfendale, A.W. (ed.): *Progress in Cosmology* (Reidel, Dordrecht 1982)
79. Zeilik, M.: *The Evolving Universe* (Harper and Row, New York 1982)
80. Zeldovich, Ya.B., Novikov, I.D.: *Relativistic Astrophysics* (University of Chicago Press, Chicago, Vol. 1, 1971; Vol. 2, 1983). Two volumes. Very good advanced book, rich in subjects.

Some Recent Articles of General Interest:

Barrow, J.D.: Cosmology and elementary particles. Fundamentals of Cosmic Physics **8**, 83 (1983). A review article containing some most recent developments in cosmology.
Barrow, J.D., Tipler, F.J.: *The Anthropic Cosmological Principle* (Oxford University Press, New York 1986)
Carr, B.J., Rees, M.J.: The anthropic principle and the structure of the physical world. Nature **278**, 605 (1979)
Chandrasekhar, S.: The general theory of relativity. Why it is probably the most beautiful of all existing theories. J. Astrophys. Astron. **5**, 3 (1984)
Davies, P.: The anthropic principle and the early universe. Mercury **10**, 67 (1981)
Dolgov, A.D., Zeldovich, Ya.B.: Cosmology and elementary particles. Rev. Mod. Phys. **53**, 1 (1981)
Dyson, F.J.: Time without end. Physics and biology in an open universe. Rev. Mod. Phys. **51**, 447 (1979)
Georgi, H., Glashow, S.L.: Unified theory of elementary particle forces. Phys. Today **33** (No 9), 30 (1980)
Hawking, S.W.: "The Boundary Conditions of the Universe", in *Astrophysical Cosmology*, ed. by Brück, H.A., Coyne, G.V., Longair, M.S. (Pontificia Academia Scientiarum, Citta del Vaticano 1982)
Hawking, S.W.: The goal of theoretical physics. CERN Courier, Jan.-Feb. 1981, p. 3, and March 1981, p. 71
Helfand, D.: Superclusters and the large scale structure of the universe. Phys. Today **36**, (No 10), 17 (1983)
Iliopoulos, J.: Les particules charmées. La Recherche **10**, 476 (1979)

Kazanas, D.: Dynamics of the universe and spontaneous symmetry breaking. Astrophys. J. **241**, L59 (1980). (The first modern exposition of an inflationary universe.)

Linde, A.: The present status of the inflationary universe scenario. Comments Astrophys. **10**, 229 (1985)

Nanopoulos, D.: The inflationary universe. Comments Astrophys. **10**, 219 (1985)

Papagiannis, M.: "Could We be the Only Advanced Technological Civilization in Our Galaxy?", in *Origin of Life*, ed. by Noda, H., (Center of Academic Publications, Japan 1978) p. 575

Penrose, R.: "Singularities and Time Asymmetry", in *General Relativity*, ed. by Hawking, S.W., Israel, W. (Cambridge University Press, Cambridge 1979) p. 58

Quigg, C.: Elementary particles and forces. Sci. Am. **252**, 64 (April 1985)

Rees, M.: Our universe and others. Quart. J. Roy. Astron. Soc. **22**, 109 (1981)

Rees, M.: "Galaxy formation", in *Kosmologie und relativistische Astrophysik*. Mitt. A.G. **58**, 57 (1983)

Schramm, D.N: The early universe and high energy physics. Phys. Today **36** (No. 4), 27 (1983)

Steigman, G.: Observational tests of antimatter cosmologies. Ann. Rev. Astron. Astrophys. **14**, 339 (1976)

Turner, M.S., Schramm, D.N.: Cosmology and elementary particle physics. Phys. Today **32** (No 9), 42 (1979)

Weinberg, S.: The decay of the proton. Sci. Am. **244**, 52 (June 1981)

Zeldovich, Ya.B., Einasto, J., Shandarin, S.F.: Giant voids in the universe. Nature **300**, 407 (1982)

Further References for the Figures

Fig. 1.3   Pananides, N. (1973): *Introductory Astronomy* (Addison-Wesley, Reading, MA) p. 248

Fig. 1.12   Forman, W., Jones, C., Cominsky, L., Julien, P., Murray, S., Peters, G., Tananbaum, H., Giacconi, R. (1978): Astrophys. J. Suppl. **38**, 357

Fig. 2.13   Mathewson, D.S., van der Kruit, P.C., Brouw, W.N. (1972): Astron. Astrophys. **17**, 468

Fig. 2.15   Moffet, A.T. (1975): In *Galaxies and the Universe*, ed. by A. Sandage (Chicago University Press, Chicago, IL) p. 212

Fig. 3.1   Silk, S. (1980): *The Big Bang* (W.H. Freeman, New York) p. 182

Fig. 3.2   Rowan-Robinson, M. (1977): *Cosmology* (Clarendon Press, Oxford) p. 54

Fig. 3.4   Seldner, M., Siebers, B.L., Groth, E.J., Peebles, P.J.E. (1977): Astron. J. **82**, 249

Fig. 5.9   Misner, C.W., Thorne, K.S., Wheeler, J.A. (1973): *Gravitation* (W.H. Freeman, New York) p. 908

Fig. 6.7   Audouze, J., Vauclair, S. (1980): *An Introduction to Nuclear Astrophysics* (Reidel, Dordrecht) p. 119

Name Index

Subject Index